Once a Wolf

Once a Wolf

The Science Behind Our Dogs' Astonishing Genetic Evolution

Bryan Sykes

Liveright Publishing Corporation
A Division of W. W. Norton & Company
Independent Publishers Since 1923
New York • London

First American Edition 2019

Originally published in the United Kingdom under the title THE WOLF WITHIN:
The Astonishing Evolution of the Wolf into Man's Best Friend

Printed in the United States of America

For information about special discounts for bulk purchases, please contact
W. W. Norton Special Sales at specialsales@wwnorton.com or 800-233-4830

Manufacturing by LSC Communications, Harrisonburg
Production manager: Anna Oler

Library of Congress Cataloging-in-Publication Data

Names: Sykes, Bryan, author.
Title: Once a wolf : the science behind our dogs' astonishing genetic evolution / Bryan Sykes.
Other titles: Wolf within
Description: First American edition. | New York : Liveright Publishing Corporation, a
Division of W.W. Norton & Company, 2019. | Reprint of: The wolf within : astonishing
evolution of the wolf into man's best friend. London : William Collins, 2018. | Includes
bibliographical references and index.
Identifiers: LCCN 2018055788 | ISBN 9781631493799 (hardcover)
Subjects: LCSH: Dogs—Evolution. | Dogs—History. | Human-animal relationships. |
Wolves—Evolution.
Classification: LCC SF422.5 .S935 2019 | DDC 636.7—dc23
LC record available at https://lccn.loc.gov/2018055788

Liveright Publishing Corporation, 500 Fifth Avenue, New York, N.Y. 10110
www.wwnorton.com

W. W. Norton & Company Ltd., 15 Carlisle Street, London W1D 3BS

1 2 3 4 5 6 7 8 9 0

To Sergio and Ulla

Contents

Preface

This book is about how wolves became dogs. A remarkable transition, it ranks as one of the most important yet least appreciated events in the long history of not one but two species. The wolf changed from a highly successful and independent carnivore into a highly successful yet completely dependent vassal with a bewildering array of different forms. The second species is, of course, ourselves.

All the evidence, which we will examine in this book, traces the start of the transition to about 40,000 years ago somewhere in Eastern Europe. Wolves had been living there and in all of the world's circumpolar regions for millions of years. Our *Homo sapiens* ancestors were much more recent players, having newly arrived from Africa only a few tens of thousands of years ago. The scene was set for the encounter that changed the world.

The location was a steep-sided river gorge in the Carpathian Mountains in what is now Romania. There is abundant evidence of human occupation in the region from the time of the Neanderthals to the arrival of our *Homo sapiens* ancestors, and there is a good fossil record of the fauna to colour in the details.[1]

I hardly need add that the narrative of this meeting, found in chapter 1, is embellished with a generous helping of my own imagination, which I hesitated to include until I read *Man Meets Dog by* Konrad Lorenz, the

Nobel Prize-winning biologist. He imagined a similar scene, though in a different location and with different players.[2] I hope you find it evocative.

In 2009 the charismatic actor, Mickey Rourke, was nominated for an Academy Award and won a Golden Globe for his portrayal of over-the-hill fighter Randy 'The Ram' Robinson attempting to make a comeback in the film *The Wrestler*. The striking parallel between Rourke, the fading actor, and his character, so it's said, was the reason behind the popularity of his nominations. In an interview with television host Barbara Walters to coincide with the film's release, Rourke said of his own past:

> I sort of self-destructed and everything came out about fourteen years ago or so … the wife had left, the career was over, the money was not an ounce. The dogs were there when no one else was there.

Asked by Walters if he had considered suicide, he responded:

> Yeah, I didn't want to be here, but I didn't want to kill myself. I just wanted to push a button and disappear … I think I hadn't left the house for four or five months, and I was sitting in the closet, sleeping in the closet for some reason. I was in a bad place, and I just remember I was thinking, 'Oh, man, if I do this,' [and] then I looked at my dog, Beau Jack, and he made a sound, like a little almost human sound. I don't have kids. The dogs became everything to me. The dog was looking at me going, 'Who's going to take care of me?'

There are tens of thousands of stories like this. Of grown men, and women, lost in the world, who are saved by their dogs.

I am a scientist, a geneticist whose research has concentrated on the human past and our own evolution from upright ape to master of the universe, or so we like to think. It was a natural step for me to wonder at the equally remarkable parallel evolution of the dog that has been so closely tied to our own.

However, and it is best that I come clean right from the start, I am not a 'dog person'. I lay the blame for this unfortunate disposition squarely on the muscled shoulders of the 'Hound of the Baskervilles', a huge Boxer living down the road from my childhood home in south-east London. From the age of seven, my route to school took me unavoidably past its house, and every day without fail the huge beast flew at the gate, ears flat back on its enormous head, snarling and gnashing its teeth. It was as if the Hell Hound itself had materialised in the London suburbs.

Many decades later, when it was suggested I write a book about the evolution of dogs, the memory of the hound came flooding back. 'I can't possibly,' I answered feebly. But as the weeks passed and I began to do a little research I realised just how fascinating a subject it was and how extraordinary is the everyday sight of a person walking his or her dog. Here was a highly evolved primate and a savage carnivore, whose ancestors were once mortal enemies, living side by side as if it were the most natural thing in the world. My re-education has only gone so far, so please, dear reader, don't expect childhood recollections of playful puppies racing across sunlit beaches or heart-wrenching accounts of how, had it not been for little Bella, I would have been unable to get over the loss of my favourite aunt. My starting point does at least allow me to be objective, even though I feel a little uneasy in being the only author of a dog book that I have come across who is not hopelessly in love with them.

Once a Wolf is primarily a book about the evolution of dogs and the forces that drove this astonishing transformation from a fierce and wild carnivore to the huge range of comparatively docile animals that is the domesticated dog. It is also about the other side of the equation, how it was that our own species *Homo sapiens*, an equally aggressive carnivore, formed such a special relationship with what, on the face of it, is a most unlikely ally. *Once a Wolf* contends that this is more than just a story of the subjugation of one species by another but a shining example of the co-evolution of two species to each other's mutual benefit. Indeed, I conclude that this co-evolution was one of the vital steps in helping *Homo sapiens* gain the upper hand in the competition with other human species, such as Neanderthals, and to expand in numbers from relative obscurity towards the overwhelming numerical superiority and influence that we enjoy today.

The scientific substance of the book draws on the rich detail of the genomes of both dog and human that has accumulated over the past two decades. Thanks to these advances we are able to make out clear patterns in the distant origins of both species, resolving questions that have puzzled scientists for over two centuries. I also explore the history and practice of breeding and its influence on the health and the welfare of pedigree dogs. In parallel, I explore the breadth of this 'special relationship' between man and dog, including interviews with the owners of many different breeds, as well as the lengths some will go to immortalise their favourite pet through cloning.

As I mentioned a few pages back, we think nothing of seeing a dog and its owner walking together along the street, yet how did this everyday scene ever come about? We have long suspected that dogs descend from wolves. We know that the distant ancestors of today's dogs formed close bonds with us a long time ago and there is a multitude of theories to account for our compatible social organisations. To a geneticist like myself, none of these is anywhere near enough to explain this most peculiar situation. In the harsh world of natural selection, only advantageous traits are conserved from one generation to the next.

Many owners who were interviewed for this book are fulsome in praise of their dog's loyalty and companionship. That may well be true today, but it is grossly inadequate to explain the rise of the dog at a time in our evolution when we were living on the edge of starvation with no time for luxuries. No, there must have been a compelling evolutionary advantage in keeping a dog, not least to offset the extra demands of feeding it.

There is another question that requires an answer. Domestication (a wholly inaccurate phrase in my opinion but which will do for now) occurred at a time when all humans were both hunters and gatherers, but mainly hunters. In this respect their way of life had not changed a great deal for at least 200,000 years. There were plenty of wolves, hyenas, jackals and foxes about which could have formed the ancestral stock for the dog – and yet there is no evidence of 'domestication' until 50,000 years ago at the outside.

Many theories seek to explain what it was that propelled *Homo sapiens* from a scarce, medium-sized primate to the position of complete domination that we enjoy today. The ability to control fire, the evolution of language and the invention of agriculture are three prominent examples. I would add a fourth: the transformation of the wolf into the multi-purpose helpmate and companion that is the dog. We owe our survival to the dog. And they owe theirs to us.

1
Lupa

At its narrowest point, the mighty Danube thunders through a narrow gorge, the Gate of Trajan,* cut by the river into the limestone bastions of the Carpathian Alps. Lupa, the she-wolf, stood at the edge of the gorge gazing down at the small figures making their way up the banks of the river a hundred metres below. They did not provoke in her any particular reaction. Humans had been using the river in this way for as long as she could remember. She and her pack did not have anything to do with the humans, but all the same she liked to keep an eye on them when they were in her territory. She knew the humans as brave hunters but they moved far too slowly to be very effective. They would eat anything that moved, including her fellow wolves if they could catch one. But that very rarely happened, and only if a wolf was sick or injured in some way. Earlier in the year, Lupa had watched the humans ambush and kill a young mammoth by driving it over the edge of the cliff, though this was unusual and most of the time they seemed barely able to scrape a living. For Lupa the main thing was to leave them alone and avoid unnecessary confrontations.

As the river mist lifted with the first rays of morning sun, Lupa could see the humans more clearly and, with her acute awareness of every detail

---

* Named after Roman Emperor Trajan (ruled 98–117 CE) and marking the northern boundary of the Empire.

of her surroundings, she sensed that they were a bit different from usual. They were a little taller, a little slimmer perhaps and moved a little more, how would she put it, a little more *gracefully*. Probably nothing in it, she thought to herself. Even so, I'll keep a close eye on them. She turned away and trotted effortlessly back across the undulating grassland, dusted by an early frost, to join the rest of the pack. It was October and winter was well on the way. The river had begun to freeze over and the last of the reindeer had already moved down from the high plateau to their wintering grounds on the river estuary. It was time for Lupa and her pack to follow them, and next day she led them on the long trek downstream towards the Great Black Sea.

Along with Lupa and her mate of two seasons there were four young wolves in Lupa's pack, two from this year's litter and two from the year before. The pups, born in June, were just old enough to learn to hunt. Before that the pack was too small to be viable for long and it had been hard work getting enough food over the summer. As always, it was Lupa who organised the hunting. She decided what prey to target, even which animal to go for. She planned the chase to take advantage of any variation in the contours of the landscape and decided where to set any ambushes. The pack was completely dependent on her skill and leadership.

Meanwhile, the humans at the bottom of the gorge were not aware that they were being watched. They knew about wolves, of course. They occasionally came across one in the forests and were familiar with the eerie howling that kept pack members in touch with one another. But in general humans and wolves kept themselves to themselves. The new type of human, *Homo sapiens*, that Lupa had seen from her vantage point at the lip of the gorge had other things on their mind. The first of these was that the gorge was also home to Neanderthals. They were noticeably different in appearance, being much heavier set and therefore stronger, but at the same time were less agile. Neanderthals and moderns tolerated each other and, in fact, occasionally interbred. The biggest difference between the two human species was invisible. The Neanderthals were not as smart or inventive. They hadn't changed their hunting methods or equipment for at least 200,000 years and showed little sign of ever doing so. The moderns on the other hand were always thinking of new ways of doing things. New designs of stone tools, of bows and arrows, the invention of the atlatl, or spear-thrower, and of all sorts of personal adornments. In time, these improvements would spell the end of the Neanderthals, and now there was one other innovation that was about to make an impact, a coalition between wolf and human, something the Neanderthals had never even contemplated.

The caves lining the Gate of Trajan were a favourite hibernation site for one of the most feared animals of the Upper Palaeolithic, the cave bear *Ursus spelaeus*, half as big again as the brown bear and with a voracious, omnivorous appetite for food which, from time to time, included humans, both Neanderthal and modern. Whereas Neanderthals abandoned the shelter of the caves as soon as they heard or smelled a bear nosing around, moderns had learned to leave the caves in the autumn and return a few weeks later when the bears were hibernating and kill them where they slept. This gave them vacant possession and enough meat to help them through the winter, should they wish to stay.

By early March the days were getting longer, although not appreciably warmer, and Lupa knew it was time to make a start for the high ground. The wolf pack had survived the winter by feeding off the herds of reindeer and wild horse which overwintered on the delta. But first there was the business of mating. Lupa was only receptive to the alpha male for five days every year. That was enough for her to get pregnant once again. She wanted to be sure to reach her traditional denning site in the hills in good time for the birth of her cubs. Very early one morning, with the frost decorating the dried stems of last year's reeds, she led her pack away from the delta and headed west for the mountains.

In past seasons Lupa had arrived in the gorge ahead of the Neanderthals, who had also spent the winter on lower ground. This year she was surprised to see humans were already living around the gorge when she arrived with the other wolves. She made her way to her usual birthing den in a small cave hidden behind a patch of eroded scree high up on the side of the gorge. Ten days before the cubs were due, she settled down and waited for the births. For the period of her confinement the alpha male ran the pack. All the wolves brought food to Lupa which they left outside her den.

In due course Lupa gave birth to four blind cubs. One, the weakest, died almost immediately, but the other three developed quickly. Their eyes opened at two weeks and a week later they were beginning to feed on regurgitated meat. The following week, Lupa led her pups outside the den for the first time where they played under her supervision. The other wolves who had kept Lupa supplied with meat during her confinement now began to do their share of babysitting, giving Lupa a well-deserved break.

The first thing she did was to walk to her favourite lookout at the edge of the gorge to see what the humans were up to. She could see a small group paddling in the river, overturning stones and occasionally plunging their hands into the icy water to pull out a crayfish. This is something the Neanderthals never did. But the biggest surprise was still to come. On her way back to the den she saw not far away on the plateau a group of humans who appeared to be hunting. The Neanderthals never came up to the top of the gorge. These strange new humans were the same slimmer version she had seen the year before. Unsure what to make of them, she kept low to the ground out of sight behind a clump of dwarf willow.

Over the rest of the summer Lupa and her pack saw more and more of the humans up on the plateau.

She saw them ambush a wild horse they had deliberately separated from the herd. They had it cornered in a patch of marshy ground below a low bluff where it became trapped in the mud. Two of the humans – there were six in all – climbed the bluff with spears in hand. While the others spread their arms and shouted to confine the horse and prevent it from escaping, the two on the bluff raised their spears and hurled them into the struggling animal. It shuddered and dropped to the ground. All six humans crowded round the stricken beast and drove their spears deep into its chest. Once it was dead they took out stone knives, opened the abdomen and shared the liver between them. They then butchered the rest of the carcass and made their way back down the gorge. Not all their hunts were as successful as this, and more than once over the summer Lupa watched as the exhausted humans made their way home empty-handed.

The first flurries of winter snow fell on the high plateau in August and the reindeer were once again on the move to lower ground. The first snows heralded the best month's hunting of the year for the wolves. Calves born in May were now almost fully grown but were inexperienced. The wolves knew which routes the animals would take across the undulating plateau and planned to intercept them in the pockets of soggy ground that lay in their path. Lupa led her pack, now nine strong, towards the ambush zone, many kilometres from their home near the top of the gorge. But something was troubling her. She stopped and sniffed the air. There it was again, the same scent she had first encountered at the site where the humans had killed and butchered the wild horse a few weeks earlier. Not only was Lupa's olfactory sense very acute, she was also able to remember smells for months or even years. She knew very well the pungent scent of the Neanderthals, but this was certainly different, still strong but a little sweeter. Scent always being her primary sense, from now on she would recognise the new humans using her nose rather than her eyes. She scanned the horizon. She could not see any humans. She led her pack onwards.

Suddenly from a small clump of birch trees about twenty metres away an enormous bull aurochs charged out, heading straight for Lupa. These giant beasts, the ancestors of domestic cattle, had very short tempers and were extremely aggressive towards wolves. Lone bulls like this one were worst of all. Wolves knew better than to take on an enraged aurochs. It would take a much bigger pack than Lupa's to subdue and kill such a giant. Before she had time to organise the rest of the pack, the beast was on her. She just managed to dodge the deadly horns on the first pass and moved backwards out of range. Seeing her in trouble, the first instinct of the rest of the pack was to protect its leader. The alpha male rushed into the attack, attempting to sink his long canine teeth into the beast's huge neck. With a flick of the bull's head the wolf was skewered on the aurochs's left horn. Another flick and the bloodied body was flung to the ground. The other wolves went to attack, still desperate to protect their leader. The thrashing bull caught one of this year's cubs full in the chest with its back leg then turned and trampled the winded and mewling animal and left it dying on the moss. Lupa herself now joined in, knowing full well that if she was killed or injured the pack was finished.

Just then, two humans appeared downwind over the crest of a low hill. They had been tracking the aurochs. They had heard the commotion and now they saw the reason for it. Standing well back, they took up position and hurled their spears at the snorting bull. The sharpened flint tips found their mark. One spear struck the animal in the flank while another buried itself deep in the beast's chest, its razor-sharp edge severing the aorta. Blood spurted from the wound and the beast fell to its knees. It lay there quivering and within a few minutes it was dead.

The two humans advanced on the carcass, knives at the ready. They looked up, expecting the wolves to retreat, but instead they held their ground and lay watching in silence. The hunters opened up the animal and removed the steaming entrails. They cut slices from the warm liver and began to eat. When they had taken their fill but before they started to butcher the carcass, the younger of the two hesitated. He had seen how wolves ran down their prey, following them for many kilometres until the animals, weak from exhaustion, could fend them off no longer. Once they were sure the death throes no longer put them in danger of serious injury, the wolves would engulf the dying animal, ripping at the exposed abdomen and disembowelling it. An idea was beginning to form in the mind of the hunter.

Reaching into the ribcage of the fallen aurochs, the younger man ripped out its still-beating heart and tossed it towards the wolves, much to the dismay of his older companion. Still the wolves stayed where they were, their amber eyes fixed on the humans. After a full five minutes Lupa was the first to move, gingerly advancing towards the offered heart. The other wolves watched in silence. Lupa sniffed at the heart, then opened her wide jaws and sliced off a chunk of the left ventricle and began to eat it. Still the others did nothing. After a further five minutes, with an almost imperceptible movement of her ears Lupa sent a silent signal to the rest of her pack. They advanced and tore the rest of the heart to shreds.

When both wolves and humans had gorged themselves on the beast's entrails they sat there looking at each other. Something passed between them. Was it a spirit message? Was it merely mutual admiration between hunters? Did either of them know what had just happened?

Over the years that followed, wolf and human grew closer together. The next spring, as lines of reindeer moved towards the skyline through purple meadows of crocus and gentian on their way to summer pastures, wolf and human followed to pick off the stragglers. Increasingly easy in each other's company, they no longer kept their distance and it was not long before they began to cooperate in the hunt. Sensing a weakness among the reindeer, Lupa picked out the target animal in the herd. The pack trotted off in pursuit, with the humans following as best they could. As the isolated deer began to tire, the wolves formed a circle and held it at bay until the humans arrived to kill it with their spears. Because the wolves no longer needed to completely exhaust the animal in order to

An artist's recreation of what a collaborative hunt, like Lupa's, might have looked like. The wolves harry the aurochs, tiring it out, while the humans inflict the killing wounds from a safe distance.

avoid injury, the chase was over more quickly. For their part, the humans had a static target for their spears. All shared the kill.

Wolf and human benefited from this collaborative hunting, and in the years that followed, long after Lupa had died, both groups learned to adapt and improve it. Wolves began to signal the presence of prey with a low-pitched howl. Humans understood the message and a hunting party set out to join them. Wolves and humans who hunted together prospered at the expense of those who did not. Their numbers increased and gradually the unstoppable current of natural selection spread this symbiosis across the rest of Europe. Eventually some wolves began to live with humans, intermittently at first, then permanently. Their numbers increased even more and, from this beginning, dogs began to evolve.

All this happened a very long time ago in the high and wild country above the Gate of Trajan. That was the start. We have yet to reach the end.

2
Darwin's Dilemma

It's easy to pinpoint the moment when the collective view of how humans and all other animals and plants came to be changed abruptly. On 24 November 1859 the naturalist Charles Darwin published *On the Origin of Species by Means of Natural Selection*. The main contention of the book, that species were not fixed and could change over time, immediately challenged the predominant view of the Church that all of nature was deliberately and carefully designed by God himself. Humans were created by God in His image and, as such, occupied a special place above all other animals. The fact that all naturalists at the two predominant British universities, Oxford and Cambridge, were enrolled as Church of England clergymen as a condition of their employment only strengthened the grip that this 'natural theology' had on scientific opinion. To disagree was dangerously close to heresy.

At the heart of Darwin's theory of natural selection was the concept that individuals within a species differed in their ability to survive and reproduce. Those that succeeded in what he referred to as 'the struggle for survival' passed on these qualities to their offspring, who were then better able to endure the struggle. Consequently, over time, new species evolved and others became extinct.

In many ways, Darwin was unlike any modern biologist. He knew nothing of genetics, the underlying principles of which lay undiscovered until well after his death in 1882. Nor did he work in a laboratory. Instead he relied on extensive correspondence with hundreds of his contemporaries throughout the world, persuading them to pass on information and sometimes to examine or collect specimens on his behalf. By these means, his accumulated wisdom and knowledge were immensely broad, which is what makes his writings such a joy to read. His theory of evolution took decades of development and refinement. Most of these were spent collecting a wide range of examples of his theory in action until he finally felt ready to publish.

One important strand was Darwin's observations of the creation of new forms by deliberate breeding, which he referred to as artificial selection. His favourite examples were the extravagant strains of domestic pigeon created by fanciers, the main reason being that he was pretty

The Expression of the Emotions in Man and Animals by Charles Darwin was published in 1872. Darwin's book is among the most enduring contributions to nineteenth-century psychology and a testament to his fascination with the dog. The left illustration is captioned, 'Half-bred Shepherd dog approaching another dog with hostile intentions'. The right, 'The same caressing his master'. Both were drawn by A. May.

certain that they all descended from just one wild species, the rock dove *Columba livia*. As in all his work, Darwin was thorough and meticulous. He kept the main varieties of pigeon himself at home, and through his network of contacts collected as many skins as he was able from far and wide. He spent days in the collections at the British Museum and even enrolled in two London pigeon-fanciers' clubs.

As well as pigeons, Darwin studied pigs, cattle, sheep, goats, horses and asses, domestic rabbits, chickens, turkeys and ducks, even goldfish, not to mention plants of many kinds. And, importantly for us, dogs. The first chapter of his accumulated thoughts on evolution through artificial selection, published in 1868 as *The Variation of Animals and Plants under Domestication*, is devoted entirely to dogs.

Right at the start Darwin sets out the principal question surrounding the origin of dogs.

> The first and chief point of interest in this chapter is whether the numerous domesticated varieties of the dog have descended from a single wild species or from several. Some authors believe that all have descended from the wolf, or from the jackal or from an unknown and extinct species. Others again believe, and this of late has been the favourite tenet, that they have descended from several species extinct and recent, more or less commingled together.

Then he adds: 'We shall probably never be able to ascertain their origin with certainty.'

Darwin's questions on the origin of dogs remained unanswered for over 120 years until the new science of molecular genetics began to take an interest. In the chapters that follow we will explore what this new science has to say about the evolution of dogs and how, for once, Darwin has been proved wrong. We *have* been able to ascertain the origin of dogs with certainty.

3

I Met a Traveller from an Antique Land

This is not the first time that I have hijacked this line from Shelley's 'Ozymandias'. It conveys perfectly the sense of antiquity and timeless continuity I still feel when I gaze at my favourite guide to the past – mitochondrial DNA.

To explain, we need to go back thirty years to a key paper published in the leading scientific journal *Nature* by the New Zealand-born evolutionary biologist Allan Wilson from the University of California, Berkeley.[1] Wilson and his team had taken placenta or cell lines from 147 women from all over the world and isolated DNA from the mitochondria. Mitochondria are components of our cells that reside in the cytoplasm, that part of the cell that surrounds the cell nucleus but is still contained within the cell membrane. They are integral components of the cell, but they have their own separate origin. Back in the distant past they were free-living algae that became engulfed by a primitive cell and have remained there ever since. Being originally separate organisms, mitochondria still retain their own DNA. Their special property is that they enable the cell to use oxygen to burn food. Until then, cells only had the apparatus for anaerobic metabolism and could not cope with atmospheric oxygen. With the help of their newly acquired mitochondria, however, cells could squeeze up to nine times as much energy from the same amount of food. In the early atmosphere, oxygen was toxic but mitochondria turned it into the life-giving gas upon which every animal species depends upon today.

The other unusual feature of mitochondria is that they are inherited only through the female line. The reason is that animal eggs are crammed full of mitochondria, while sperm don't have any to speak of. To be entirely accurate, those few they do have don't survive in the fertilised egg. This was the feature that appealed to Wilson and his team. Everyone inherits their mitochondrial DNA from their mother, who got it from her mother, who inherited it from her mother and so on back through time. Males and females have mitochondrial DNA – after all they both need to breathe oxygen – but only females pass theirs on to their offspring.

In complete contrast to mitochondria, the DNA in the cell nucleus is inherited more or less equally from both parents. This nuclear DNA controls most of the body's functions, with the important exception of aerobic metabolism, which remains the responsibility of the mitochondria and its DNA. Unfortunately, ancestral connections traced backwards by nuclear DNA soon become extremely complicated. We all have two parents, four grandparents, eight great-grandparents, sixteen great-great-grandparents and so on. The number of ancestors doubles with each past generation, so by the time we go back only twenty generations, that's about four hundred years for humans, we have over a million ancestors. It's very unlikely that we have inherited DNA from all of those thanks to the random mixing of nuclear DNA with each generation, something I shall explain later in the book. Even so, we will have inherited DNA from a great many of them, but from whom we will never know. In comparison with this genetic muddle, there was only ever one woman in each generation who is our mitochondrial ancestor and whose DNA we have inherited. It is that simplicity which drew Allan Wilson to investigate mitochondrial DNA (or mDNA for short) rather than nuclear DNA in his representative sample of the world's population.

The striking conclusion of this work was that if you went back far enough, everyone on the planet has inherited his or her mDNA from just one woman. In ways that we will come on to, Wilson estimated that she lived in Africa about 200,000 years ago. Unsurprisingly, she was immediately dubbed 'Mitochondrial Eve'. The results also showed a clear connection between Africans and everyone else, suggesting that modern humans spent a long time in Africa before some of them left to populate the rest of the world. It's as well to remind ourselves here that we are only considering strict female–female matrilineal inheritance, with no consideration, for now, given to the DNA from men.

It was a delightfully simple conclusion, although some people still find it confusing. Eve was certainly not the only woman alive at the time, just the only one to have direct matrilineal descendants living today. As now, couples can have only sons or no children at all, but it is only daughters who can pass mitochondrial DNA to the next generation. It follows that in the 10,000 or so generations since Eve, the only mDNA to survive to the present day has been passed along unbroken matrilineal lines, while that from Eve's many contemporaries has been eliminated at some point along the way.

Though there have been some modifications in the ensuing thirty years, this overall concept of Mitochondrial Eve has stood the test of time. Wilson's 1987 paper became a model for all future molecular genealogies, which have completely revolutionised our view of human origins. I analyse mitochondrial DNA samples from all over the world, and marvel at every one of them. They have each travelled unseen for tens of thousands of years in the cells of a continuous line of ancestors from ancient times until today when, at last, they reveal their secrets in the laboratory.

It took ten years before the Los Angeles-based biologists Robert Wayne and Carles Vilà published an equivalent genetic analysis for the dog.[2] Like Wilson, they used mitochondrial DNA, but with a more advanced technique that examined the DNA sequence itself rather than the limited summary that was all that had been available to Wilson a decade earlier. I will say more later on about DNA sequences, including what they are and how to read them, but for now we will concentrate on the dogs.

Wayne and his team collected an impressive set of samples. In addition to 140 domestic dogs from 67 different breeds, Wayne also included wolves, coyotes and jackals in his analyses. The wolf collection came to a total of 162 animals from 27 locations worldwide. In addition, because they had been mooted as possible ancestors of modern dogs, Wayne included 5 coyotes and 12 jackals – 2 golden, 2 black-backed and 8 simien. When the mDNA sequences from all these animals were displayed in a molecular tree (referred to as Wayne's tree) in the same way that Wilson had portrayed the human mitochondrial genealogy, the resemblance between the two was clear to see.

Wilson's human tree (see top diagram, page 17) divided the world population into two main branches, one African and a second containing both some African and all the people from outside Africa before coalescing on a single matrilineal ancestor – 'Mitochondrial Eve'. The Wayne dog DNA tree consisted of four main branches, each with a different, but still closely related, ancestor. Most dog breeds were placed in the major branch, which Wayne called branch I, and included many of the common breeds as well as some so-called 'ancient breeds' like the dingo, New Guinea Singing Dog, African Basenji and Greyhound. Branch II contained two Scandinavian breeds, the Elkhound and the Jämthund, while branch III included a variety of breeds such as the German Shepherd, Siberian Husky and Mexican Hairless. Finally, branch IV included Wirehaired Dachshund, a Flat-coated Retriever and an Otter Hound. This last branch also contained a few wolves, one of which, from Romania, was the only wolf in the whole study whose sequence exactly matched that of a number of dogs including a Toy Poodle, a Bulldog and, surprisingly, another Mexican Hairless.

The simplified diagram (see bottom of page) only shows the major mitochondrial groups. Within each circle are a number of breeds. They are not shown here but can be inspected in the original,[3] where there are many examples of exactly the same mDNA sequence being found in several different breeds. For example, a Norwegian Buhund, a Border Collie and a Chow Chow had precisely the same mitochondrial DNA sequence. Equally, the same breeds could have different mDNA sequences and appear on different branches of the tree. For instance, the eight German Shepherds had five different sequences between them. We will consider what this means a little later.

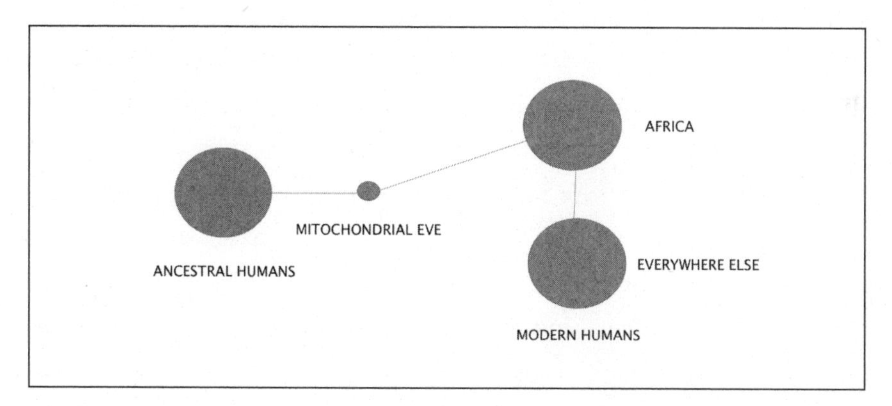

Wilson's Human Tree (simplified).

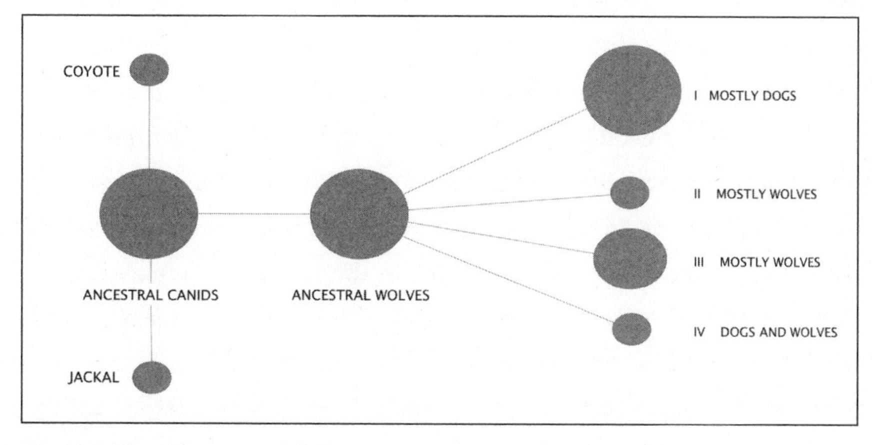

Wayne's Dog Tree (simplified).

Had Darwin been alive to read it, he would have been itching to know where wolves, coyotes and jackals fitted into the tree, if at all. The answer was very clear. The coyote and jackal fell out of the main wolf/dog tree, as it were, immediately. Their DNA sequences were clearly quite different to all the dogs, and none made it into any of the four major branches. When it came to placing the wolf DNA sequences, the answer was equally striking, not because they were outside the dog tree but because they were deeply embedded within it. There was no doubt, from the mitochondrial DNA analysis, that all dogs were descended from wolves and from no other species. It was the first triumph of molecular genetics as applied to dogs – and by no means the last.

Darwin wasn't wrong about much, but, by means he could never have foretold, his statement on dogs that 'We shall probably never be able to ascertain their origin with certainty' would turn out to be one of those rare exceptions. I am sure he would have been utterly delighted to be proved wrong.

Turn the clock forward another ten years to the present day and the Wayne dog tree is still alive. But, like the technical improvements we have considered in the decade between the Wilson and Wayne papers, there have been great strides in DNA analysis in the last ten years, which have led to some radical pruning of the original tree, while leaving the major branches intact.

Before we turn to the effect of these improvements in filling in the blanks in our knowledge of dog evolution there is one other important genetic system to consider. This is the Y-chromosome, the mirror image of mDNA in a genealogical sense in that it traces not the maternal but the paternal genealogy through time. Again the reason is simple enough. Only males have Y-chromosomes and they pass them on exclusively to their male offspring. In many species it is a less reliable witness than the mitochondrial equivalent because of the very variable mating success of males. In most species, including our own, males have the potential to father virtually unlimited numbers of offspring, or none at all, but females are restricted to just a few. This has major implications when we come to look at pedigree dogs.

Just as any conclusion about evolution based on mitochondria should carry the caveat that it can only reveal patterns based on females, so the Y-chromosome only traces the origins of males. Of course, ultimately they both have to tell more or less the same story, but there are fascinating twists and turns along the way.

In any sort of genetic analysis it is vital to be able to detect inherited variation, which is the lifeblood of genetics. Variation comes in many different forms – blood groups, hair colour, height or DNA sequence. You simply can't do any genetics without it. For DNA the variation in sequence can be read directly, as it is for most mDNA comparisons, or it can use what are known as genetic markers. These are places where the sequence between, in this case, different Y-chromosomes, is known to differ. You can then test for the markers directly without having to sequence the whole chromosome, which saves a lot of time and money. But before you can use them you have to find them, which used to be an enormous bore. It is much better now, as we shall see.

The tedious process of discovering dog and wolf Y-chromosome markers was slow to get going and the first studies used a panel of only four markers. Luckily Y-chromosomes alone are spared the process of shuffling with other chromosomes, something else I will explain as we go along, and so the markers can be combined as blocks. So four markers (A–D), with two versions at each one (1 or 2), can differentiate sixteen Y-chromosomes (A1, B2, C1, D2; A2, B1, C2, D1 and so on), meaning that you can do a lot with just four markers and sixteen combinations.

A group from Sweden was the first to publish any wolf and dog results from this kind of analysis, having studied both Y-chromosomes and mitochondria in 314 dogs from 109 different pedigree breeds.[4] Their wolves came from six different regions in Europe and North America, a total of

112 animals. And of course, for both dogs and wolves, all the animals were males. It came as no great surprise after the Wayne mitochondrial pattern (as illustrated on page 17) to find that the dog and wolf Y-chromosomes were similar. Also, there was no sign of any other species, as was always a formal possibility when only the mDNA results were known. Had the original dogs been hybrids between female wolves and male jackals, for example, this would have been invisible to mDNA analysis but not to that of the Y-chromosome. The confidence that wolves really were the only ancestors of all dogs increased substantially after the Swedish study.

In a similar fashion to mDNA, the same Y-chromosome, as defined by its genetic markers, was to be found in several different breeds of dog. As an example, an identical Y-chromosome was found in a Bernese mountain dog, a Border Collie, a Dalmatian, a Greyhound, a Poodle, a Shetland sheepdog and a West Highland terrier. On the other hand, different individual dogs of the same breed often had several different Y-chromosomes. Five Collies, for example, were found to have three different Y-chromosomes between them.

The comparison of the male and female genetic contributions showed quite clearly that in domestic dogs there were many more different mDNA sequences around than there were different Y-chromosomes. What that meant became clear when the wolf results were compared. In wolves the number of different mDNA and Y-chromosome sequences was about the same, not skewed as in dogs. This is very familiar scenario in many human populations where there are lots of different mDNA types but fewer Y-chromosomes than there should be if breeding success was roughly equal across the sexes.

Wolves are almost entirely monogamous, with only one breeding male and one breeding female in a pack. As a consequence males and females make an equal overall genetic contribution to successive generations and, as the Swedish team found, this balances the mDNA and Y-chromosome diversity. In pedigree dogs, the situation is more like some human populations where a few males have a disproportionate number of offspring. The ultimate human example is Genghis Khan, the thirteenth-century Mongol emperor who has an estimated 16 million male descendants living today, each of them carrying his Y-chromosome. Genghis Khan achieved this feat by slaughtering his male enemies defeated in battle and inseminating as many women as possible, often to the point of exhaustion. 'Try spending the night alone from time to time,' his doctors cautioned. When he died in 1127 Genghis passed on his wealth, and his habits, to his sons. Male dogs can achieve similar breeding success with considerably less effort than Genghis Khan. All they have to do is win 'Best in Show' and let the breeders do the rest.

At first it was a puzzle as to why pedigree breeds showed little or no sign of a common origin, at least as far as mDNA and Y-chromosomes were concerned. Surely with all the care taken to make sure pedigree dogs breed true, all dogs within a breed should have the same origins along both male and female ancestral lines. Not so. Instead, there seemed to be no telling, short of DNA testing, to which mDNA or Y-chromosome branch any particular dog belonged. Certainly, the breed could not be predicted from the DNA results from either system.

Although the scope of mDNA and the Y-chromosome is limited to just two genetic systems, we must not make the mistake of underestimating the importance of mitochondrial DNA and the Y-chromosome in crashing through the barriers of uncertainty surrounding the origin of the dog. Scientists from Darwin onwards have pondered this question with no means of coming to a definite conclusion. Were jackals or bush dogs or wolves or coyotes or foxes or hyenas or some other animal, possibly long extinct, the true ancestors of the modern dog? The research, first with mitochondrial DNA and then with the Y-chromosome, has made the answer crystal-clear. Wolves, and only wolves, are without question the ancestors of all living dogs.

4
On the Origin of Wolves

The genetic trees drawn with the help of mitochondrial DNA and the Y-chromosome make it very clear that the only ancestor of all dogs is the wolf. There is more we can decipher from the DNA results, but before we come to that, what do we know about the ancestry of wolves?

The Age of the Mammals began in the Cretaceous period following the sudden extinction of the dinosaurs some 65 million years ago. This extinction left a big gap in the fauna which was gradually filled by mammals, which until then had been an inconspicuous group of small furry animals cowering in the undergrowth. Their numbers increased, and by 40 million years ago, during the Eocene epoch, the emerging mammals began to evolve into today's familiar groups: horses, deer, elephants, apes, dogs and cats, early forms of modern Orders.

Wolves, and therefore dogs, belong to the last of these Orders, the *Carnivora*. It has become the most diverse of any Order, embracing over 280 species, and it takes its name from the Latin that describes the main characteristic of its members: 'flesh-eating'. They are the carnivorans, as distinct from the general term 'carnivores', which takes in all meat-eating species, be they fish, reptiles or even plants. A major division of the Order *Carnivora* is the Family *Canidae*, which includes wolves, coyotes, jackals and foxes. Cats large and small belong to the Family *Felidae*, bears and pandas to the *Ursidae*, while badgers, hyenas and seals belong to other, separate, families. Some carnivorans, like the Giant Panda, are strictly vegetarian but are still included within the same Order. It is their teeth that set the *Carnivora* apart from other mammalian Orders. All have well-developed third incisors, which serve to pierce the flesh of prey animals to prevent escape and to kill, while unique to carnivorans are their fearsome carnassial teeth, taking the place of our molars. Carnassial teeth are razor sharp and self-sharpening and are designed to slice through flesh like a pair of shears rather than merely tearing at it.

The dog-like carnivorans, the *Canidae*, and the cat-like *Felidae* began to diverge from each other and become gradually more specialised. As far as we can tell from the fossil record, the earliest canids evolved in North America where the oldest fossil dog, *Cynodesmus*, was discovered in Nebraska, USA, and lived between 33 and 26 million years ago. At one metre in length, it resembled a modern coyote and, by its dentition, was clearly carnivorous with large canine teeth for grasping and tearing the flesh of its prey. Soon after, on an evolutionary timescale, some truly fear-some carnivores began to evolve, including the bone-crushing *Cynarctus*. As these monsters became extinct about 11 million years ago they were replaced in turn by other canids, notably *Tomarctus*, found all over North America from Florida, north to Montana, west to California and south to Panama. From the size of the jaw muscle insertions in their skulls it is clear that *Tomarctus* had a bite strength far greater than required to kill its prey. This led to the conclusion that, like modern-day hyenas, *Tomarctus* was able to crush bones to reach the nutritious marrow of scavenged carcasses.

By the middle of the Miocene epoch, some 10 million years ago, the canids had spread from America, first to Asia, then to Europe and finally to Africa. As they did so, the ancestors of today's wolves gradually evolved towards a lighter, faster frame in order to hunt swift herding prey like elk and wild horse. They hunted not as individuals but as members of a pack. Thus began the key development in the evolution of the modern wolf,

5 The Living Fossil

All our efforts to reconstruct the past can only ever give an approximation of what really happened. Well-preserved fossils are spectacular but rare and their discovery can only ever convey a patchy record. History is notoriously inaccurate, depending on the inclinations of the author. Mythologies require sophisticated interpretation. Genetics is no different. It is just another foggy lens through which we try to make sense of times gone by. Bearing that in mind, let us clean the eyepiece and take another look.

In Chapter 3 we saw how DNA from living dogs and wolves was able to reconstruct a plausible genetic relationship between the two. These were inferences from modern DNA but, astonishingly, DNA can survive for thousands of years in fossil bone and teeth. As we will see later, it is often in a pretty bad state. Nonetheless it does give us the chance to examine ancient sequences directly rather by inference. Later on, we will have a closer look at how ancient DNA has helped us follow the evolution of the dog. But before that we need to know a little more about DNA itself.

All genetics depends on mutation, the ultimate source of all variation. DNA changes over time. When a cell divides, its DNA is copied so that each of the two daughter cells contains the full set of genetic instructions. The copying process is astonishingly precise and accurate, and the error rate is minuscule. Editing mechanisms within each cell scan the copies for errors and correct them. But the error rate is not zero. After each cell division, roughly 1 in 1,000 million mutations gets through uncorrected. If the emerging mutation changes a vital component of a gene, then the daughter cell will either malfunction or die. Only extremely rarely will a mutation be beneficial. The most dangerous malfunctions are those that turn normal cells into malignant ones which lose the capacity to restrain their own cell divisions, and they develop into tumours. That is why in some rare diseases, where the DNA editing and correction capacity of cells is faulty, it leads to much higher rates of malignancy.

Fortunately, the majority of DNA copying errors has no consequences whatsoever; first, because the errors don't occur in important genes, or, second, because they are not passed to the next generation. Only mutations in the germ line, being the cells that go on to form eggs and sperm, are capable of travelling on through time. Even then, the vast majority of sperm never get to fertilise an egg, and, in mammals, most eggs are not fertilised anyway. For these reasons alone, the overwhelming majority of germ-line mutations which occur through faulty copying, even the most potentially damaging of them, are not passed on.

Some mutations, however, do get through to the next generation. Most will not be noticed and have no significant effect on the body, either because they occur in unimportant genes or in the gaps between genes in the long stretches of DNA whose function, if any, is still largely unknown. Here it is worth distinguishing between genes and the rest of our DNA. Genes do something, usually instructing cells how to make proteins. There will be more on this when we take a look at gene mutations that have been found in dogs, but for now we will concentrate on the inconsequential mutations that have no effect, neither good nor bad. Precisely because they are so inconsequential, these unassuming 'neutral' mutations are the lifeblood of the sort of genetic reconstructions of past events that we have covered so far. A damaging mutation in a vital gene will disadvantage the individual who carries it. Not necessarily fatally, of course, but enough to put him or her at a slight reproductive disadvantage and thus reduce the prospects of the mutation being passed on to the next generation. Generally, over time, the mutant gene will be eliminated by selection, though not always, as we shall see in pedigree dogs. However, the humble and meaningless mutations that have no effect on anything of great importance will escape the scrutiny of selection and will sail on unmolested through future generations. It is these humble mutations that are the guiding lights that illuminate the history written in the language of the genes.

To explain how mutations are used, in dogs and humans, to date past events like the timing of the transformation from wolf to dog, let us imagine a desert island in the middle of a vast ocean. A young couple arrives in a canoe. For our purposes, it could equally well be a couple of shipwrecked dogs. The island is a paradise, with plentiful fresh water in bubbling streams flowing down from high mountains in the interior. There are coconut palms, shellfish and crabs in the sea and no predators or dangerous animals to disturb the idyll. Everything needed for life is on hand, and the couple start a family. Their children grow up in this cradle of abundance and, ignoring incest taboos for the sake of this exercise, have children with their siblings.

Time passes. The population settles down to a stable total of 1,000. Nobody leaves the island, there is plenty for everybody, and no one else arrives. Until one day a scientist and a research assistant turn up and begin taking a DNA sample from each of the inhabitants. The samples go off to the lab and the sequences are read. A few weeks later the scientist and his assistant, now back home, get the results. What can they deduce about the people on the island from the results? It doesn't matter all that much which genetic markers we are talking about for this example to work, so let's keep it simple and imagine that we are working with mitochondrial DNA. The first things the researchers notice is that everyone's DNA sequence is very similar. Some sequences are identical, and we will call that the 'core' sequence of the island. However, about half the people have a sequence that differs at just one DNA base from the core.

DNA sequences are written using a childishly simple alphabet with only four letters. These letters represent simple organic chemicals, or *bases*, joined together in a linear sequence. Their abbreviations are even simpler : A, G, C and T. Any DNA sequence is a long string of these bases: ... CCGGTAA ... and so on. A mutation might change a T to an A, making the new sequence read as ... CCGGAAA ... The language may be child's play, but the meaning is far from simple, as we will explore further in a later chapter, but not now. Instead we travel back to our island.

Of the 1,000 people who were tested, 500 have the core sequence and 500 have a one-base difference from it, but not all at the same one.

The researchers draw the reasonable conclusion that everybody on the island is ultimately descended from one couple, or rather from one woman, as we are dealing with mitochondrial DNA. Can they tell from the results how long ago the island was settled? To get an answer, we need to agree a very important factor. The mutation rate. That is the rate at which mDNA mutations occur and get passed on. It is going to be an estimate, drawn from other results. The factors which contribute to the estimate are sometimes astonishingly crude.

A common approach is to take two species, say human and chimpanzee, compare their DNA sequences and make an assumption about how long it is since they last shared a common ancestor. In this example the usual figure is 6 million years, based on fossil evidence which, for both species, is extremely flimsy. All genetic dating of past events depends crucially on the accuracy of the mutation rate and that it has remained stable over the period.

Fortunately, the estimates of the mitochondrial DNA mutation rate by the various methods come up with a figure that most are happy to accept. For the segment of mitochondrial DNA that Wayne and Vilà used, the rate is estimated to be one base change every 20,000 years. Mutations occur randomly as cells divide, so we must turn to discussing probabilities. A mutation rate of one per 20,000 years doesn't mean that no mutations occur until that time has passed. It is an average. It could happen in the first generation or the last or, more likely, somewhere in between. Let us say the time between generations on the island is twenty years. If a quarter of the population has a mitochondrial DNA sequence that is one mutation away from the core, the average number of mutations per person across the whole island is then one quarter. The estimated time from first settlement then becomes the average number of mutations (a qaurter) from the core, multiplied by the mutation rate (20,000), which comes to 5,000 years.

Returning to our scientists, they go back to the island to inform the council of elders of their results of the project. They also reveal that the original settlers had come from the mainland far away to the east because that is where they have also found the core mDNA sequence among the inhabitants. After listening politely to the presentation, the elders turn to the scientists and, as I have experienced first-hand, they say, most politely, something like 'Thank you for your trouble. We knew that all along.'

In my deliberately simplified example we were dealing with just one segment of DNA on an island originally settled by only one couple. No one arrived or left for millennia. It doesn't get any simpler than that.

Let us now suppose that other things happened on the island. Perhaps half the population died in an earthquake, or the central volcano erupted, destroying the crops, and three-quarters of the people starved, or an epidemic killed 90 per cent of the population. These are the sorts of catastrophes which might have happened in real life. Those events can severely distort our calculations. For instance, and let's make it extreme, a tsunami kills everybody on the island except a couple who were far out to sea fishing at the time. They survived and, over time, their offspring repopulated the island. In this scenario, the genetic calculations would give the time that had elapsed since the tsunami rather than since the original settlement. The island would have undergone a 'population bottleneck'. There would be no way of telling, by genetics alone, for how long the island had been settled before the tsunami struck. If we introduce further complexity, like a few boatloads of new arrivals, then all hope of being precise about the original settlement date goes up in a puff of smoke.

Given these unknown and often unknowable factors, I take claims of accurate genetic dating of past events with a large pinch of salt. That does not mean they lack value, but it is a mistake to become a slave to such calculations. We will use the island metaphor again when we come to consider the origins of pedigree dog breeds. Wayne and Vilà also used this kind of calculation to estimate the timing of the wolf–dog transition. The answer was much further back than anyone suspected, between 76,000 and 135,000 years ago.

6 Let the Bones Speak

At some point in the past the lives of wolf and human became intertwined and it is from this partnership that the dog eventually emerged. Until genetics entered the fray, the only way of following this transition through the intervening millennia was through fossils. Good fossils are in short supply and the fossil record is understandably full of gaps.

In terms of time, the oldest skulls that could even remotely be differentiated from wolves were excavated in the Goyet cave in southern Belgium in the 1860s. Like all good fossil sites, Goyet is a limestone cave whose alkaline environment helps to preserve the calcified bones and, importantly, any DNA that might lie within.

From studying the style of stone and bone tools found there, it was clear that the cave had been occupied by humans for a very long time. Neanderthals lived there during the time of the Mousterian culture, which lasted from about 160,000 years to 40,000 years BP (the standard archaeological abbreviation for 'before present'). It takes its name from the rock shelter at Le Moustier in the Dordogne region of central France. The Mousterian lasted until the arrival of modern humans, our ancestors, about 40,000 years ago. As is not uncommon with early excavations, disturbance of the layers within the cave made precise stratigraphic dating of the different artefacts found there problematic. However, carbon-dating of the fossils gave precise dates for the organic remains at least. The cave fauna was a rich assemblage of cave bear, cave lion, horse, reindeer, lynx, red deer and mammoth. In the deeper recesses of the cave archaeologists found the skull of a 'large canid' carbon-dated to 31,700 years BP. Was it a wolf or was it a dog?

Of course, there must have been a period after the first wolf was adopted into a human band when its skull was exactly the same as a wolf's – because it was a wolf. There was no exact moment of transition from one to the other, and the whole debate has a strong flavour of semantics. The more cautious authors merely refer to these intermediates as 'canids' or 'wolf-dogs', thereby sidestepping the argument altogether.

A similar conundrum faced archaeologists excavating the nearby site of Trou des Nutons, a cave formed in the limestone hills of the Ardennes by the River Lesse, a tributary of the Meuse. Among the fossils found in the Trou des Nutons were beaver, roe deer, horse, bison and wild sheep, suggesting a later occupation than at Goyet. This was confirmed when another skull of a mystery 'large canid' was given a carbon date of 21,800 years BP. This is a surprisingly early date and in the middle of the last Ice Age. But was it the skull of a dog or a wolf?

These skulls from France were subjected to a series of precise measurements of snout-length and width, the length of the tooth row and the size of the flesh-shearing, self-sharpening carnassial teeth that wolves and dogs have where we have molars.

Fossil canid skulls from two archaeological sites in Russia and Ukraine, one at Mezin (Ukraine) and the other at Avdeevo just over the Russian border, were given the same treatment. These two sites were inhabited by early humans who constructed huts of mammoth bones and left behind an abundance of beads and other artefacts carved from mammoth ivory. The objective of the osteometric study of candid fossils from these two sites was to discover whether the remains of these 'large canids' differed sufficiently from wolves in their skull morphology to be classified as dogs on their way to domestication rather than unmodified wolves.

To complete the comparisons, the analysis was extended to include later, but still prehistoric, unambiguous fossil dogs from France and Germany. Also included were a selection of modern and fossil wolves from Europe and Asia along with modern dogs from several large breeds including Great Dane, Tibetan Mastiff, Siberian Husky, Chow Chow, Irish Wolfhound, Malinois, Dobermann Pinscher and German Shepherd.[1]

Comparing multiple skull measurements from dogs of different sizes is a complicated business, and I will spare you the details of the multivariate analysis and go straight to the main conclusion. The Palaeolithic skulls from the oldest sites, including Goyet at 31,700 years BP, had a significantly different shape from modern, or indeed fossil, wolves. This suggests that, even by that early date, these animals were dogs already on the way to modification through 'domestication'. An alternative explanation, though in my opinion rather less likely, is that these were the skulls of one or more wolf species that later became extinct. As we shall see later, there is other enticing evidence to support the former scenario and suggest that the close association between wolf and man began a very long time ago.

The next layer of evidence about the changing appearance of domesticated dogs comes from the late glacial period around 17,000 years BP, when the ice sheets covering northern Europe were fast retreating. The shrinking tundra no longer supported herds of large prey animals. The climate warmed considerably, rainfall increased and forests covered much of the formerly open tundra. The fauna changed with the landscape and many prey animals disappeared. Mammoths, woolly rhinoceros and their predators, the sabre-tooth tiger and cave bear, were forced into extinction. Others, like the wild horse, reindeer and bison, shifted their ranges. Humans began to spread north, first following the shrinking herds and later, as they entered the Mesolithic period, changing their diet to smaller woodland prey, like wild boar, pine marten, red and roe deer. On the coastal settlements, shellfish became a major source of food and the first boats ventured out to sea to catch fish. Supplementing this meagre protein diet were roots and tubers, insects and snails. The heroics of the mammoth hunt became a thing of the past and life became a gruelling fight for survival.

The close cooperation between human and dogs, by now thoroughly assimilated into human society, continued even though the superbly effective working partnership that had developed in the Upper Palaeolithic was at its best when killing large prey, a practice which by now was rapidly disappearing.

Around 12,000 years BP much smaller dogs made their debut in the fossil record. A team of French archaeologists found the remains of thirty-nine dogs at the Pont d'Ambron rock-shelter in the Dordogne. From an osteometric analysis similar to that carried out at the earlier sites of Goyet and Trou des Nutons in the Ardennes, it was clear that the Pont d'Ambron dogs were considerably smaller. The same was true with the remains excavated at the Montespan cave in the northern foothills of the Pyrenees and at the open-air site of Le Closeau in an old channel of the River Seine.

The authors of the exhaustive paper summarising this body of work confidently concluded that they were dealing with the remains of dogs and not wolves. In France at least, and also in Spain, dogs were clearly changing. In Russia, however, at around the same time, wolf-dogs were still very large. Whether this was a result of separate wolf domestications in the two regions or for some other reason, it was impossible to say. One firm but rather grisly conclusion, drawn from cut-marks on the bones of the Pont d'Ambron dogs, was that they had been butchered and, presumably, cooked and eaten.

As well as the issue of timing, the identification of the geographical location of the wolf–dog transition has absorbed many researchers and continues to do so. The first scenario to be proposed, by a group from the University of Konstanz in Switzerland led by Peter Savolainen, was that the major 'domestication' event happened only once, in East Asia.[2] This was the conclusion of an mDNA study of 654 dogs from different regions of the world where the focus was on the diversity of sequences. The perfectly sensible rationale was that the highest diversity, that is the highest number of different mDNA lineages, would be found in the places where dogs had been around the longest and had the most time to accumulate new mutations, rather like the islanders in our metaphorical example. Savolainen's team found mDNA sequence diversity was highest in south-east Asia and located the first 'domestication' to the region. This was a very controversial conclusion at the time, and it would be another decade before the debate was settled, although it still rumbles on in some quarters.[3]

In order to make progress on the vexing issues of timing and location, scientists turned to the DNA that had, incredibly (a word I do not use lightly), survived in fossils. Robert Wayne, who headed the Los Angeles lab, was one of the eclectic bunch of scientists who dared to think, against all reason and common sense, that DNA might survive in fossils. As there was no academic tradition of ancient DNA science and this was an entirely new field, the early pioneers came from all sorts of backgrounds. Svante Pääbo, for example, who went on to sequence Neanderthal DNA, was originally an immunologist with an interest in Egyptology that led him to attempt to extract DNA from mummies in 1985. Ed Golenberg, who claimed in a 1990 *Nature* article that he had extracted DNA from a 17-million-year-old magnolia leaf, was a botanist. Scott Woodward, in a paper published by *Science* in 1994, reported DNA extraction from a fossil dinosaur *Tyrannosaurus rex* from the Cretaceous period entombed in a block of coal. Woodward was a geneticist from Brigham Young University in Utah who went on to run a large genetic genealogy project for the Mormon Church. My own background was in medical genetics, specifically the causes of inherited bone disease. In 1989 my colleagues and I reported the first recovery of ancient bone DNA in *Nature*.

We met regularly to feel our way in this exciting but tricky field where extravagant claims could be accepted for publication by the very best journals – and, more often than not, be rapidly dismissed. Robert Wayne was a regular attendee at these meetings. He is an evolutionary zoologist with an interest, at the time, in the hybridisation of wolves and coyotes where their ranges overlapped. Robert has gone on to become the pre-eminent scientist in dog genetics, first with work on fossil DNA and then with extensive analyses of the genetic variation in living dog breeds. Much of what we know about the genetics of dog evolution comes from Wayne's lab in Los Angeles. I was slightly surprised to discover that Wayne doesn't own a dog, but he does have a cat.

Once the field settled down in the years following the initial papers on ancient DNA recovery, a number of labs began to report its successful extraction from fossil wolves and unambiguous dogs, sometimes of great antiquity.

The field advanced in fits and starts, at first with the publication of single cases, then a few related finds and eventually, in 2013, a large series that seems, for now, to have settled the question of the origin of the wolf–dog transition in favour of Europe between 19,000 and 32,000 years ago.[4]

In the first decade of this century, the protocols for recovering ancient DNA improved a great deal and it became realistic routinely to obtain long sequences from old bone. Once again mitochondrial DNA was the target, for the very good reason that there are far more copies in a cell compared to nuclear DNA. If you are working at the limits, as you always are with ancient DNA, you want to make things as easy for yourself as possible.

DNA sequencing technology had also advanced to a point where it became practicable to sequence all 16,727 bases of the canid mitochondrial genome from fossils. Analysing the complete sequence avoided the potential bias of restricting the analysis to the shorter 'control region' used in the earlier papers by Wayne and Vilà and by Savolainen. The large 2013 study used more or less complete mitochondrial sequences of eighteen fossil 'canids' along with a large collection of modern dog breeds. Although not every specimen yielded all base pairs of sequences, it was enough to place them accurately on the evolutionary tree. Nuclear DNA, conversely, was too badly preserved to be of much use.

The resulting tree, or phylogram, to use the proper name, again recognised the four main branches (I–IV in the figure on page 17) of modern dog breeds initially published by Wayne and Vilà. The results were fascinating. The fossil dogs on three of the four branches (I, III, IV) of the tree are closely related to modern breeds while the rare fourth, mainly Scandinavian, branch (II) is closest to modern wolves from Sweden and Ukraine. One possible explanation is that dogs on this branch, which include the Norwegian Elkhound and the Jämthund, acquired their mitochondrial DNA from wild wolves in the recent past, after the advent of agriculture.

While all of the ancient dog lineages have survived to the present day, that is not the case for the fossil wolves. Many of these lineages are now extinct or have simply not been picked up in living wolves yet, though the likelihood of that diminishes as more and more modern wolves are sequenced.

There is a wealth of fascinating detail in the 2013 paper by Olaf Thalmann, which I encourage you contemplate at your leisure from the original publication.[5] I do, however, want to mention one particularly surprising finding – about dogs in America. Only two fossil dogs were sequenced, one from Argentina and the other from Illinois, USA. From these mitochondrial sequences these dogs were clearly both related to branch I European dogs, though the ages of the fossils (1,000 and 8,500 years BP respectively) mean that they must have arrived well before the first European settlement in the fifteenth century. These dogs accompanied the indigenous Native Americans who had arrived earlier from Asia.

None, however, had mitochondrial DNA remotely like that from American wolves. This has to mean that Native American dogs were ultimately descended from European and not American wolves.

There was another surprise in store. Breeds thought to have been descended from indigenous 'Pre-Columbian' dogs, like the Chihuahua and Mexican Hairless, also had an exclusively European mitochondrial heritage. Although sample numbers are quite low, it does look as if the indigenous Native American mitochondrial lineages were another casualty of European settlement.

As the dust settles on the controversies still hovering over the timing and location of the transition from wolf to dog, one thing is certain. It all began a very long time ago.

7 The Cave of Forgotten Dreams

Though hardly fixing the dawn of the transition between wolf and dog with any degree of precision, the genetic dates are embedded in the bounds of what we call the Upper Palaeolithic – the last of the three phases of the Old Stone Age.

The origin of this classification, which is still used today, can be traced to John Lubbock, 1st Baron Avebury. A banker by profession, he also had a wide range of other interests including politics, biology and archaeology. His interest in the natural world grew from his friendship with Charles Darwin, who moved to the same village, Downe, in Kent, in 1842 when Lubbock was eight years old. As Lubbock matured, his interest in evolution and archaeology grew. He became an ardent supporter of Darwin's evolutionary theories and of academic liberalism in general. He bought land in Wiltshire to save the famous prehistoric stone circle at Avebury from destruction and introduced into Parliament a bill that would eventually become the Ancient Monuments Act, the forerunner of all legislation to protect ancient sites.

Lubbock divided the Stone Age into two phases, the Palaeolithic, sometimes known as the Old Stone Age, lasting until roughly 10,000 years ago, and the Neolithic, the New Stone Age which followed it, coinciding with the invention of agriculture. Later an intermediate phase, the Mesolithic or Middle Stone Age, was adopted as the term for the period between the end of the last Ice Age about 17,000 years BP and the dawn of agriculture when the Neolithic began. About 4,000 years ago, the Neolithic gave way to the Bronze and then Iron Ages. The Palaeolithic was further divided into Lower, Middle and Upper phases, with the last of these lasting from about 50,000 years BP until the transition to the Mesolithic. Incidentally, the dates here only apply to the Stone Age in Europe. In other parts of the world the transitions occurred more recently; indeed, in highland New Guinea the Stone Age lasted until well into the twentieth century.

The genetic dating places the wolf–dog transition firmly within the Upper Palaeolithic, a quite extraordinary period in the history of our species, bristling with innovation and new ideas. The hallmark of the Upper Palaeolithic is the appearance of new forms of stone tools, the most durable of evidence. Until then, the only tools were hand axes and spear points. They were carefully made, certainly, but had not changed in basic design for tens of thousands of years. Suddenly, archaeologists were finding delicate arrow points, bone needles, even fish hooks, artefacts never seen in older, deeper layers.

Human fossils were much scarcer than stone tools, but they too showed a change from heavy-boned and robust skeletons whose skulls boasted prominent brow-ridges and receding chins to an altogether lighter and more graceful form. Was this a change brought about by slow adaptation, or was it the sign of the arrival in Europe of a new human species? After years of debate the argument was settled in favour of the wholesale replacement of the indigenous humans – *Homo neanderthalensis*, the Neanderthals – by a new arrival from Africa. This was *Homo sapiens*, our own ancestors. Mitochondrial genetics was the deciding factor in settling the argument in favour of replacement.

A few days before Christmas 1994 three speliologists, Eliette Brunel-Deschamps, Christian Heller and Jean-Marie Chauvet, were clambering over the face of the Ardèche gorge in southern France. Here the river cuts through the southern flanks of the limestone Massif Central on its way to join the Rhône near St-Just, well on its way to the Mediterranean. It is in the nature of limestone to form underground cave systems when exposed to the constant attention of slightly acidic groundwater. Over thousands of years the water gradually erodes the rock, hollowing out caverns of sometimes immense proportions.

The entrances to some 4,000 caves, many little more than overhangs, some as vast as medieval cathedrals, punctuate the steep walls of the gorge. Most are clearly visible, while others are hidden by rock-falls and vegetation. It was in order to find these hidden caverns that Jean-Marie Chauvet and his companions were inching their way along the steep sides of the gorge. They were searching for air currents emerging through cracks and crevices that would betray the presence of a cave system deep underground. At one point Chauvet felt a slight breeze coming from the rocks brushing the hairs on the back of his hands. He bent down to sniff the subterranean zephyr, then called his companions over. They agreed that the gentle flow smelled promisingly, damp, ancient and strangely alive. One by one they carefully removed the small rocks surrounding the vent until they came to a narrow cleft through which the air was escaping. It was too narrow for any of them to squeeze through, so next day they returned with hammers and a small pneumatic drill and set to work widening the crack. They found themselves faced with a narrow shaft descending into the black depths. Being experienced, not to say fearless, potholers they squeezed through the gap until they reached a point where a gallery opened out in front of them. It was clear, only from inside, that the main cave entrance had been blocked by an ancient rock-fall and that they had somehow found their way into the main cavern through the roof.*

As they explored the cave system over the following days the true wonder of their discovery began to dawn on them. Gleaming stalagmites

* This is the 'Cave of Forgotten Dreams' of the chapter title. I have taken the name from Werner Herzog's excellent 2010 documentary film about Chauvet cave.

rose up from the cave floor to meet their counterparts, delicate stalactites, suspended from the roof. Formed by the steady drip-drip-drip of calcium-rich water, their pristine condition showed that no animal or human had disturbed these hidden depths for thousands of years. On the floor lay scattered bones and skulls of cave bears petrified beneath a glassy coating of calcite.

As Chauvet and his companions pushed further and further back along the galleries they saw in front of them the first of the paintings. Dozens of crude human hand prints outlined in red ochre covered one of the walls to a height of nearly two metres. These were only an introduction to the treasures which lay further back. There, on the deliberately smoothed cave walls, were drawn the images of lions, bears, mammoths, rhinoceros, horses and giant deer. These are the oldest morphologically

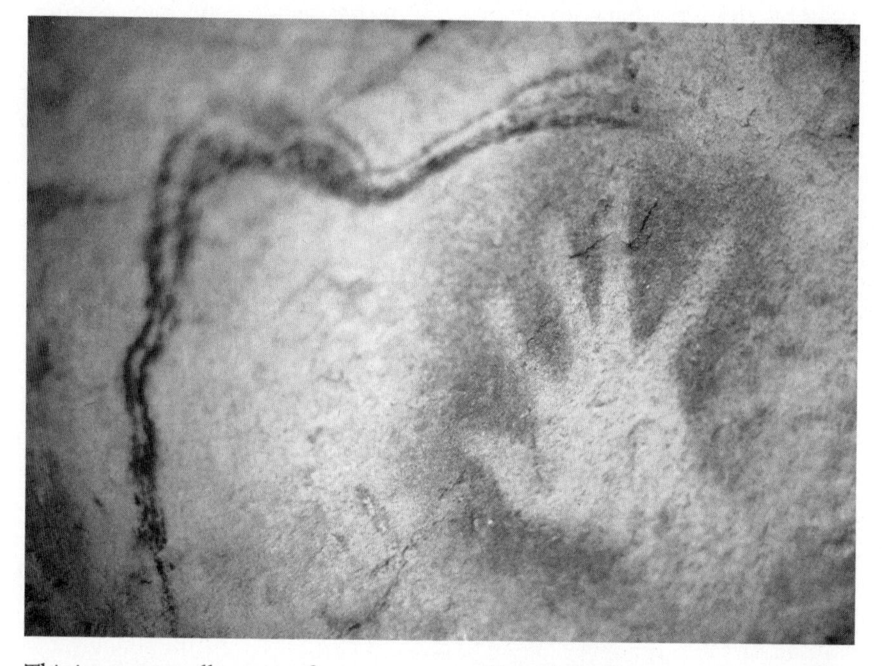

This image actually comes from an exact replica of Chauvet cave that opened in Vallon-Pont-d'Arc in 2012, as access to the ancient caves is severely restricted for the protection of the artwork. The replica art was created using the same tools and methods as it is believed were used by the original artists.

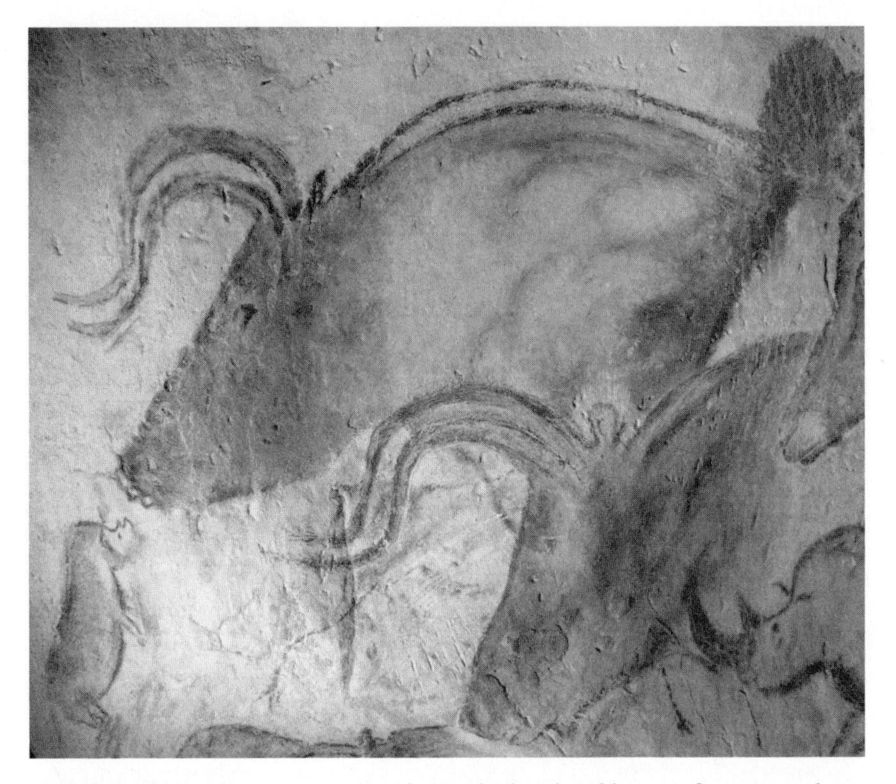

A painting from Chauvet cave that shows the head and horns of two aurochs, an extinct form of wild cattle that would have been a key prey animal for both humans and wolves.

accurate depictions anywhere in the world. What strikes home about them is their beauty. These are not merely crude outlines like the hand prints in the antechamber. They have form, expression and movement.

As well as being objects of wonder in themselves, these paintings have naturally led us to contemplate the reason they were drawn in the first place. What a task it must have been. Working deep underground without any natural light, the artists, for that is what they were, could only illuminate their lithic canvases by the light of glowing wooden torches. Streaks on the walls show where they had rubbed the dying embers to rejuvenate the flames. Carbon-dating the charcoal smeared on the walls was the means of discovering how long ago the drawings were made.

All organic material contains carbon, and this can exist in two forms called isotopes. Carbon 14 is very slightly radioactive. The other isotope, carbon 12, is not. After an animal or plant dies, or is burned in the case of the wooden torches, the radioactive carbon 14 slowly decays with a half-life of almost 5,000 years. In other words after 5,000 years there is only half as much carbon 14 remaining. By comparing the content of the two isotopes using a mass spectrometer to count the atoms, the age of the specimen can be estimated. Atmospheric carbon 14 is generated by ionising radiation from the sun high up in the atmosphere, some 32 kilometres above the ground. The proportions of the two carbon isotopes in the atmosphere are more or less in equilibrium. Thus the ratio of the carbon isotopes in a freshly dead animal or plant is equal to the atmospheric ratio at the time.

There are many factors that can change this ratio artificially and consequently introduce errors in dating. One is contamination of old material with modern carbon, for example from the archaeologists who recovered the specimen. This tends to make the material appear younger than it actually is. As is well known, carbon dioxide levels in the atmosphere have rocketed due to human activity since the Industrial Revolution. This carbon is ancient, coming as it does from the burning of fossil fuels that are millions of years old and no longer radioactive. This tends to reduce the carbon 14 in a specimen and artificially increase its apparent age. Nuclear testing also affects atmospheric carbon but in the opposite direction. Enormous amounts of carbon 14 are released into the atmosphere by a nuclear explosion, which in turn reduces the time estimate for radio-carbon dating. Nowadays these influences are incorporated into the calculations and the dates produced are referred to as 'calibrated'. The original pioneers of radiocarbon dating did not take these influences sufficiently into consideration, and as a result many of the dates claimed in the earlier days of carbon-dating are wrong.

Thankfully, Chauvet cave was not discovered until the modern era of calibrated radiocarbon dating, and the dates obtained from the charcoal and other organic material in the caves can be relied upon. They show that Chauvet cave has been used for at least 80,000 years, first by cave bears, the skulls and bones of which litter the cave floor, then by an assortment of more recent Upper Palaeolithic mammals including hyenas and a couple of wolves.

There appear to have been two distinct phases of human 'occupation'. The first was between 37,000 and 33,000 years ago and most of the drawings date to this phase. A later phase of occupation which produced the crude hand prints outlined in red ochre lasted from 31,000 to 28,000 years ago.

Chauvet cave is one of a handful of decorated caves from this remarkable and crucial phase in human evolution, the others being Lascaux in the Dordogne region of south-west France and Altamira in Calabria, northern Spain. Unlike the other two, Chauvet is in pristine condition, never having been open to any but bona fide researchers under strict instructions not to disturb the cave in any way. Altamira and Lascaux were open to the public for many years before the damaging effects of exhaled moisture and carbon were fully appreciated. They are now effectively closed to prevent further damage, though visitors can enjoy the visual impact of the caves and their paintings in nearby reconstructions.

In many people's opinion the Upper Palaeolithic warrants comparison with other transformational periods in human cultural history: the rise of democracy in ancient Greece, the Italian Renaissance, the Age of Reason. So many new things were happening to the way we lived and most importantly to our interactions with the world around us. Many of these developments remain unseen and only reveal themselves in very special circumstances. Such a one is the discovery of Chauvet cave. There must be other caverns like it still sealed inside their limestone tombs, waiting for their breath to percolate to the outside. These caves give us rare glimpses into a vanished world, so very different from our own. Yet we see from the drawings that in many ways the artists were very much like ourselves. We understand the murals. Without difficulty we sense their beauty.

There are no human remains in Chauvet cave and, other than the drawings, very little sign of human presence. Nobody lived in Chauvet. What then was the purpose of these drawings, made with such effort and such skill? Clearly they were not purely decorative in the way we might hang a favourite painting on the wall above the fireplace. Although we will never know for certain, to many eyes these beautiful drawings are a tangible expression of a world of imagination and spirituality that marked the rise of truly modern humans.

An aspiration to go beyond what is absolutely necessary for function is also apparent in the stone tools our ancestors left behind. Whereas Neanderthals made perfectly functional tools like hand axes and thrusting spear points, they appear clumsy in comparison to the beautifully fashioned arrow points of the Upper Palaeolithic. The flint itself was traded over long distances and it supplied the raw material for individual craftsmen to demonstrate their skill. Fashioning a flint arrowhead or spear point was an opportunity not just to replace equipment lost in the hunt but also to demonstrate a high level of dexterity.

Quite suddenly, archaeological sites of the period were flooded with personal adornments. Excavations in south-west France reveal the appearance of bracelets, pendants and beads exquisitely fashioned from bone, antler and ivory. Seashells from the Mediterranean are found in sites hundreds of kilometres from the coast. Splinters of stone called burins were used to drill out holes in animal skins so that they could be sewn together with sinews for clothing. The effort involved was substantial.

Further afield at Sungir, 200 kilometres to the east of Moscow, archaeologists have excavated five human burials dated to 32,000 years BP, one of which contains the remains of a boy almost covered in strands of beads. There were nearly 5,000 beads in all, each one taking an estimated forty-five minutes to an hour to produce, a total of at least 4,000 hours in the making. On his head he wore a cap decorated with more beads as well as the canine teeth of at least sixty Arctic foxes. This was evidently the resting place of someone from an important family, a clear sign of social stratification emerging very soon after our ancestors arrived in Europe.

Our impression of our Stone Age ancestors is one of brutish simpletons clinging on in the face of appalling odds, surrounded by vicious and hungry predators looking for an easy meal. Certainly, their world was full of danger and life was hard. Nevertheless not every minute was taken up by the struggle to survive, and the evidence from Sungir shows that in some circumstances there was enough of a surplus for what we might imagine to be luxury, at least for a few. This was the world into which the wolf-dog was welcomed.

Far from being frightened prey cowering inside dank refuges, by the beginning of the Upper Palaeolithic our ancestors were well on the way to becoming top predators themselves. One by one the carnivores that once struck fear into the Neanderthals were driven to extinction. That perennial whipping boy of evolution, climate change, may have been the underlying influence behind the diminishing herds of mammoth and bison. But the climate had been changing for a very long time. Only when our ancestors arrived on the scene around 40,000 to 50,000 years ago did numbers of megafauna plummet. First the mammoth and the woolly rhinoceros vanished from the steppes, followed by the giant elk *Megaloceros*, then the bison and the wild horse. These were the herbivores that nourished the guild of carnivores and whose demise presaged their own extinction. The cave lion, leopard, hyena and sabre-toothed cat all vanished. The fearsome cave bear *Ursus spelaeus* who fought our ancestors for living space soon followed. Only the brown bear, *Ursus arctos*, managed to survive in the face of human competition by more or less abandoning meat altogether and restricting its diet to plants, berries and the occasional small mammal. At the end of the Palaeolithic all the large mammals, herbivores and carnivores alike, whose images jostled for space on the lime-smoothed walls of Chauvet, were gone.

A parallel wave of extinctions swept North America once humans arrived in numbers. As a species, we have never been good at taking responsibility for the damage we inflict, and the role humans played in the extinction of the North American megafauna is hotly debated. There is no doubt in my mind that both in Europe and in America it was our own human ancestors that pushed species after species over the edge into oblivion. In Europe none survived the hunting onslaught of our ances-

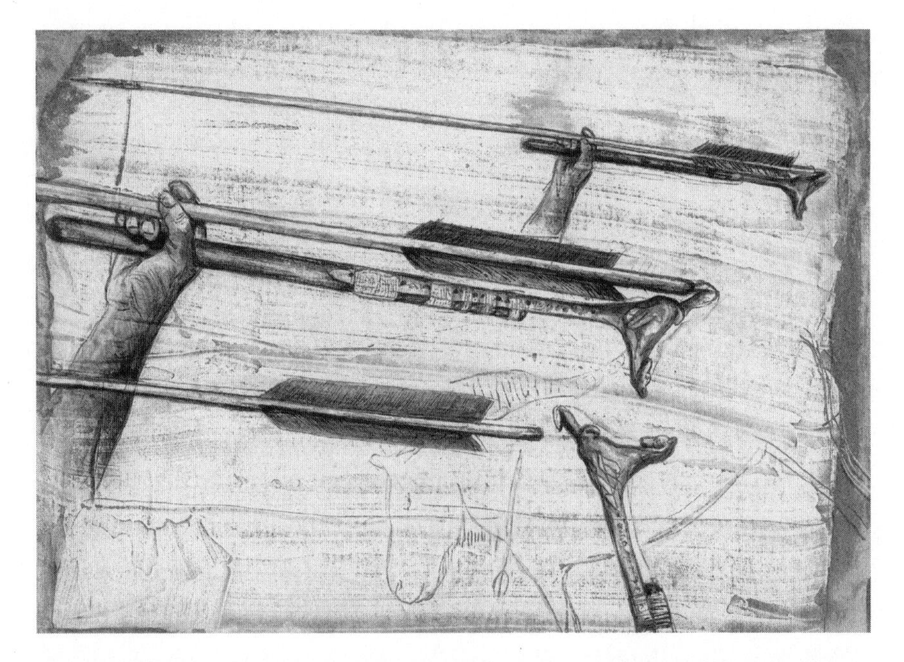

Artwork of how a spear-thrower (or atlatl) is used to throw a feathered dart. At top and centre, the dart is loaded. At bottom, it is being thrown. The angular momentum imparted means dart speeds of 150 kilometres per hour can be achieved, making this a lethal weapon in the hands of a skilled hunter. In the background is a cave painting of an animal wounded by an atlatl dart. The earliest such weapons were found at Schoningen, Germany, in the 1990s, dating to over 300,000 years ago.

tors, but in North America, when the mammoth and the woolly rhinoceros went under, the elk and the buffalo survived.

Although technical invention in hunting equipment no doubt helped our ancestors to become the top predators, there was more to it than that. Certainly, the atlatl or spear-thrower was a deadly innovation which allowed our ancestors to kill at a distance and avoid injury, though alone it was hardly enough to account for the decimation of the megafauna. This piece of equipment was certainly one ingredient in our progress towards dominance of the Upper Palaeolithic world, but it seems unlikely that we achieved this distinction just because we could throw spears further and with more force than before. It was the revolution in our minds that took place all those years ago, as witnessed by the lavishly decorated burials of the children of Sungir and the gleaming frescoes of Chauvet, that really made the difference.

In a narrow gallery towards the back of the Chauvet cave complex is a tantalising glimpse of another secret of this revolution. Impressed in the soft sediment that forms the floor of the cave are the clear impressions of a child's footprints. He or she, let us say he, was about eight years old. What he was doing that far back in the pitch-black cavern we can only imagine. From the measurements of his gait he was walking quite normally, not running nor hesitating on his way to the back of the cave. Human footprints of this age are very rare indeed, but it is not the track of the boy that makes this so unusual and important for our story. Alongside the child's is another, quite different, set of prints. Perfectly preserved in the limey sediment that covers the cave floor, hidden from view and undisturbed for 30,000 years, are the tracks of a fully-grown wolf.

We cannot be sure that the boy and the wolf were walking side by side when they made the tracks or whether the prints were left thousands of years apart. The passage is very narrow at this point and yet the tracks never cross each other, making it more than likely that they were made at the same time. Was the wolf hunting the boy? Or were they exploring the cave together, companions in a great adventure? The footprints hint at a very close relationship, friendship even, between the boy and the wolf.

Or was the animal that trotted comfortably at the boy's side no longer a wolf, but already on its way to becoming a dog?

8

Hunting with Wolves

The great changes that swept through Europe in the Upper Palaeolithic were products of the refreshed human mind, capable not only of great innovation but also of seeing the world in a different way. An essential element in this process was our ability to learn from observation, to reproduce novelties whenever we saw them. This ability is still with us today, aided by language and other forms of communication. Nowadays new ideas spread around the world almost instantaneously. Even though in the distant past the diffusion of ideas would have been much slower than this, invention and creativity were a characteristic of the time. Whether it be the best way to make an arrowhead or a spear-thrower, or the method of threading beads to make a necklace, or how to make a primitive flute by drilling holes in the wing bone of a swan, all these ideas spread through observation and duplication.

The observant Palaeolithic hunter would have seen wolves in action, relentlessly pursuing their prey until the hunted beast was exhausted, then surrounding it. Unable to inflict a fatal bite through the spinal cord like a lion, the wolf must lunge to inflict flesh wounds until, at last, the animal collapses through loss of blood. It is not a pretty sight as, more often than not, the animal is disembowelled while still alive. More to the point, the endgame is dangerous for the wolves, who risk injury from the last desperate thrashings of their dying prey.

Our ancestors would have witnessed this drawn-out death struggle and realised how easily they could have killed the cornered beast. A spear thrown from a safe distance at an animal held at bay by a wolf pack would be an easy kill. The remarkable stamina of the wolf pack could run down the swiftest prey. By comparison human hunters were no match in the pursuit, but with spears, bows and arrows they could kill even the largest animal cornered by wolves with little risk of injury to themselves.

No doubt at first the final act would seem to the wolves like yet another theft of a kill. Wolves have always been vulnerable to having their kills taken over by more dominant predators. Their ability to rapidly consume, or 'wolf down', the nourishing internal organs like the heart and liver, and then to quickly carve off huge chunks of meat using their razor-sharp carnassial teeth, was an ancient adaptation to minimise loss.

In such a situation, it is only a small step for a human hunter to realise how to pacify the wolves. Sharing the carcass is all that is needed. Cooperative hunting is of obvious benefit to both sides, if only the wolves themselves could appreciate the benefit of the deal. This approach would not work with lions or bears, but wolves hunted very much as we did, cooperatively in small groups with each member having a separate role.

There is precious little evidence to support my proposition of a working alliance between man and wolf. It makes a great deal of sense to join forces with a wolf pack in the pursuit of sustenance, so it is a reasonable and attractive speculation, but I freely admit that it is the product of my imagination. I felt nervous about proposing it until I discovered that the great zoologist Konrad Lorenz had already envisaged a similar scenario. In his charming book *Man Meets Dog*, Lorenz writes a fictional account of cooperative hunting between humans and a pack of jackals, which he saw as the wild ancestor of modern dogs.[1] He was mistaken about the jackal, as we now know, but he could equally have chosen the wolf. In 2015, the archaeologist Pat Shipman proposed that a hunting alliance between man and wolf was a major factor contributing to the extinction of the Neanderthals.[2]

I much prefer this explanation, of hunting cooperation leading to trust, for the origin of 'domestication' to the alternatives. Foremost among these, and the one most geneticists seem to prefer (even though I suspect many of them have never seen a wolf), is that wolves became accustomed to human company by hanging around their camps and picking up scraps of food from rubbish tips. As well as being dreary in the extreme, this explanation falls down simply because 'domestication' was already well under way by the time humans congregated in large enough settlements to produce sufficient waste to sustain an animal with the appetite of a wolf. Nor does it explain why, of all animals including coyotes, jackals, badgers and bears capable of surviving on refuse, none ever developed a bond of the strength and depth that comes close to matching that between human and wolf – in its modern incarnation, the dog.

Almost nothing remains of human activity on the open plains, so evidence of cooperative hunting is always going to be hard to find. Only in the dank recesses of subterranean caverns can we find physical evidence of our distant ancestry. At Chauvet it is not bones nor teeth but paintings and those enigmatic footprints that are the lens through which we glimpse the lives of our ancestors. Eight hundred kilometres north of Chauvet another river, the Samson, cuts through another limestone gorge, its walls pierced by caves. Here at Goyet, though, there is no need to search for the ancient breath of hidden galleries. The caves are wide open and unlike Chauvet have been occupied by humans, both Neanderthal and modern, for a very long time. Excavations at Goyet began in 1867, three years after the Neanderthal type specimen was excavated in the eponymous valley close to Düsseldorf in Germany.

The cave at Goyet contains large numbers of human bones from about 120,000 years BP, including Neanderthals and modern humans along with thousands of artefacts. Our interest is focused on one skull found in a crevice and dated to around 32,000 years BP, in the same age range as the first Chauvet drawings. There is no doubt that it is the skull of a canid, but whether it belonged to a wolf or a dog or something in between is uncertain and, as you might expect, fiercely debated. The snout is certainly shorter than that of a modern wolf, so it shares that characteristic with dogs. Again predictably, and rather like the case of the original Neanderthal

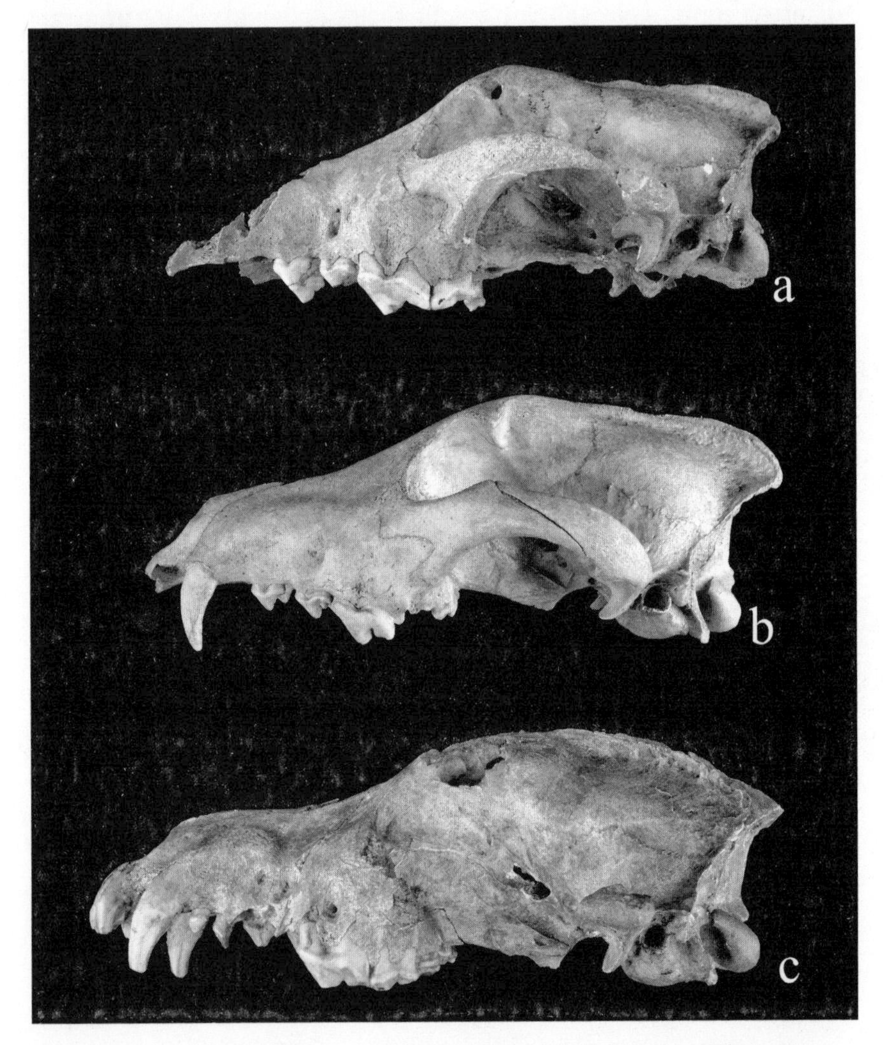

The short snout and wide braincase of a canid skull (a) found in Belgium's Goyet cave, in comparison with two ancient wolves found in nearby caves (b, c), led scientists to claim that the Goyet skeleton is from a 36,000-year-old dog. This image is reproduced courtesy of the Royal Belgian Institute of Natural Science and Wilfred Miseur.

skull, which was at first thought to be that of a deformed human, some see it as merely a wolf with a very short nose.

If wolves and humans were discovering the mutual advantages of hunting together, perhaps one might expect wolves to feature prominently in the images at Chauvet, Goyet and elsewhere. Yet they are conspicuously absent from these murals. The only wolf-like image from this area is a crude outline from the Dordogne, at Font-de-Gaume, an area rich in limestone caves with a long history of prehistoric human occupation, both Neanderthal and modern. The drawing is among over 200 images of contemporary animals, including the usual suspects like mammoth, bison and woolly rhinoceros, and dates to around 17,000 years BP. Other than this, there are no depictions of wolves at Font-de-Gaume or in any nearby caves. The other notable subject about from these cave paintings is ourselves. There are no images of humans anywhere to be seen. Why not? It is as if the taboo our ancestors felt about creating a human image also extended to the wolf.

I am reminded of Conan Doyle's short story 'The Adventure of Silver Blaze', concerning the theft of a prize racehorse and the murder of its trainer. In solving the crime, Sherlock Holmes explains to Inspector Gregory of Scotland Yard that the dog guarding the stables must have recognised the culprit.

'Is there any other point to which you would wish to draw my attention?' asks Gregory.

'To the curious incident of the dog in the night-time,' replies Holmes.

'The dog did nothing in the night-time,' responds Gregory.

'That was the curious incident,' replies Holmes.

9

Why Didn't Shaun Ellis Get Eaten by Wolves?

If there was anything in the notion of an ancient pact between man and wolf that preceded 'domestication' then the best way to find out was to meet some wolves.

In my research for this book I had read an account by a man who had managed to forge a great friendship between himself and a pack of wild wolves. Although very far removed from the Upper Palaeolithic when we, I imagined, first began to hunt with wolves, I wondered whether the basic instincts on which that relationship was based would still be there.

Shaun Ellis, the co-author of *The Man who Lives with Wolves*, describes how, as a young man, he had become instantly entranced by the sight of a captive wolf.[1] Shaun's transforming experience had happened in a wildlife park near Thetford, not far from his childhood home in Norfolk. The brief encounter changed Shaun's life entirely and led him eventually to devote himself to wolves. He moved to the USA to work as a volunteer at a wolf rehabilitation centre in Idaho and later set out on his own to find and live among a wild wolf pack.

I heard that he had returned from Idaho and, now in his early forties, was living on a farm in Cornwall. I knew I needed to meet Shaun as soon as possible and, if he would allow it, interview him. Fortunately, he agreed, and on a cold and dark Friday in December, my wife Ulla and I set off on the long train journey to Lostwithiel, where Shaun and his partner

Kim now live in the company of a small pack of wolves. We settled into the cosy parlour of the farmhouse, at least partially revived with a hot cup of tea.

I began by asking Shaun about his childhood on his grandfather's farm deep in the Norfolk countryside, leading up to that first encounter with the wolf. An only child, Shaun had been raised mainly by his grandparents, his mother having to work full-time to support the family. His grandfather in particular had been an important influence, fostering Shaun's growing love of the outdoors. They took frequent walks together with their dogs in the woods and across the fields, hunting rabbits. Always a reluctant pupil, Shaun left school at 15 and had various jobs including a spell in the army.

A free and easy man, his life continued with no definite purpose until the day he took a bus to the local zoo in Thetford. Wandering around, he came upon the wolf enclosure. There, only a few metres away, stood a fully-grown wolf looking straight at him. Here, right in front of him, was the savage killer he had been taught to fear. As he stared back into its golden yellow eyes he felt as though it was touching his very soul. It was as if the creature could read his deepest thoughts and somehow understood him better than any human had ever done. In that instant Shaun's future path through life was fixed. He has devoted himself ever since to understanding and to helping the rest of us to understand and appreciate what these wonderful creatures are really like. He was, however, sufficiently self-aware to realise that the wolf probably looked at every visitor in the same way.

No matter, over the next years Shaun's life revolved entirely around wolves. Lacking any professional qualifications and viewed with suspicion by some professional biologists, Shaun has lived with wolves, has brought up a family of wolves, and in the process he felt as though he had almost become a wolf himself. For two years he was alone with a pack of wild wolves in the forests of Idaho. He had travelled to that part of America to work at a wolf education centre where a captive pack was kept in order to educate visitors and to disabuse them of the wolf's thoroughly undeserved reputation for savagery. After a while Shaun realised that he could never *really* understand the enigmatic animal only by working with captive wolves, so he packed his rucksack and set off alone into the forest.

Shaun knew there were wolves living there – they migrated every year down from Canada and across Montana into Idaho. But he did not know exactly where they would be. Three months passed before he saw any sign of a wolf. Then, one day, he saw the unmistakable paw print of a large male in the soft mud by the side of a waterhole. In the weeks that followed he heard the chilling chorus of a howling pack drifting through the night.

Eventually he caught a glimpse of his first wolf, a large black male, as it crossed the track some hundred metres ahead. The wolf looked at him briefly before trotting off into the forest. Weeks passed and Shaun saw the black wolf more regularly as if it were sizing him up, whether as a meal or a mere curiosity he could not be certain. A few weeks later, the black wolf reappeared with four others, two males and two females – a real pack. Slowly, day by day, week by week, the wolves became more confident until one day the big male came up to Shaun, sniffed him and suddenly nipped him just below the knee.

Although it was painful, Shaun knew from his experience with captive wolves that this was not a bite in anger. Rather like shaking hands, it was a way for the wolf to introduce himself.

All wolf packs maintain a strict hierarchy. As a rule, only the dominant pair, the alpha male and female, breed, while the other wolves are given supporting roles in the pack. Generally speaking, the other wolves are the siblings and offspring of the alpha pair. Whereas the alpha female is the undisputed leader of the pack, it falls to the beta animals, one rung down in the hierarchy, to maintain discipline and to organise the defence of the pack when it is threatened. It was the beta male, the enforcer, that sized Shaun up and gave him that painful bite, not in anger but as an introduction. The encounters between Shaun and the wolves became more and more frequent over the next weeks. After six weeks he felt that he had been accepted by the pack as an 'honorary wolf'. He was given the job of what he describes as a diffuser, a calmer of nerves. Like any other family, wolf packs can argue, and they have sharp teeth and claws with which to make a point. Shaun's role, as he saw it, was to make sure these petty squabbles did not get out of hand.

'Why,' I asked him, 'were you not attacked?'

'Because I was useful. Also I suspect because the wolves could learn from me, something they needed to do, surrounded as they were by wolf-hating bloodthirsty hunters.'

Shaun's joy at being accepted by the pack was severely tested one day when, without warning, the beta male suddenly lunged at him and knocked him to the ground. The wolf stood over him, fangs bared, and growling. Here is Shaun's account of his terrifying experience.

> I suddenly felt this was the end. Up to then I had trusted them completely not to harm me, but in that instant I remember thinking OMG they *are* this creature after all. They are going to kill me. He kept pushing me back with nips and growls into a hollowed-out tree then stood outside snarling. What a fool I've been, I thought to myself, they have been biding their time "fattening me up for Christmas" so to speak. But the fatal attack never came. I was there pinned down for an hour and a half then, just as suddenly as the attack had begun the wolf backed off. He called me with a low-pitched whimper and went into a very appeasing body position as if to apologise. It was as if it had suffered some kind of psychosis, turning into a killing machine and then back again.
>
> It took me a while to come out, and that is when he led me down the path to where I was going. There by the stream were the unmistakable marks left behind by a huge bear. Not only did he save

> my life but the episode reinforced my original belief that I should always trust them.

After that, Shaun spent the remainder of his two years living peacefully with the pack.

Others like Shaun have found themselves strangely attracted to wolves and sought out their wild company. Sometimes they have written about their experiences and their books have become bestsellers. Farley Mowat was one such author.[2] Mowat was a government biologist dispatched in the 1960s to the Keewatin Barren Lands of northern Manitoba for eighteen months to report on the falling numbers of caribou and confirm, as was widely thought, that wolves were to blame for this decline.

The shortage of caribou was worrying the powerful hunting lobby that pressured ministers to do something about 'the Wolf Problem'. The 'problem' boiled down to an insistence by hunters, guides and lodge owners that wolves were destroying the big game on which their businesses depended. According to the hunting lobby, it was well known that wolves had an insatiable appetite for killing way beyond the need to feed themselves, although the blatant irony of this conviction was lost on the hunters. Mowat's mission was to investigate the accusations and report to embattled ministers.

Like Shaun Ellis, it was a long time before Mowat encountered his first wolf. Attracted by what he thought were the pathetic cries of a young dog in distress, he set out to find the source of the whining. In order not to scare what he thought was a wolf cub, he tracked the whelps from behind a gravel ridge. Peering over the crest, he found himself only two metres away from a fully-grown Arctic wolf. For a few seconds each stared into the other's eyes and Mowat fell under the same hypnotic spell that had entranced Shaun Ellis. The wolf was the first to break free from the spell's grip. He leaped into the air and, with an effortless trot, vanished into the crepuscular gloom of the Arctic dusk.

In the succeeding weeks, as Mowat got to know the wolves better, he realised that, far from being the active observer, it was he, the human, who was being studied by the wolves. More than once after vainly scanning the horizon in front of him for signs of his 'quarry' he would turn around to see two or three wolves resting on the ground a few metres away, just looking at him. Like Shaun Ellis, he was aware of being weighed up by the wolves, but for what purpose he could not be sure.

As spring turned to summer, the tundra was streaked with long lines of caribou sauntering north to their summer territory, and Mowat had a stroke of good luck. From the top of an esker, a gravel ridge deposited by melting ice, he located a den with four young cubs playing outside with no sign of the parents. A curious cub, catching his scent, began to approach when suddenly, and with a piercing howl of alarm and warning, an adult returned. Mowat lost his footing and tumbled down the side of the ridge, ending up in a tangled heap at the bottom. Half expecting the big male, the 'enforcer', to clamp his enormous jaws around his throat at any moment, Mowat turned over to look up the slope. He saw not one but three adults standing in line abreast, peering down at him with amused delight. When they had seen enough of this hilarious performance they turned and silently withdrew out of sight.

Later that night, safely back in camp, it dawned on him that all the stories he had been told of wolves as savage and demented killers were completely untrue. Earlier, these animals had the hapless Mowat completely at their mercy and yet they did nothing to threaten him, even though he was close to their own cubs. After this, his view of the mission on which he had been sent changed completely. He lost all fear of the wolves, moved out of the fetid cabin he was occupying and, leaving his guns behind, pitched his tent close to the den.

Over the following months he saw that, while wolves certainly did kill caribou, they only did so at certain times of the year. The rest of the time, for example when the caribou had moved to their own summer feeding grounds in the far north, Mowat was surprised to find that the wolves lived almost entirely on mice, lemmings and ground squirrels supplemented by the occasional pike or lamprey ambushed as they made their way up the narrow channels of the muskeg to spawn.

Many things impressed Mowat about the wolves he lived with that summer. He admired their close and amiable family life, their ingenuity at finding food, and their tolerance of his presence. What impressed him most of all was their ability to communicate with other packs over long distances. One day towards the end of the summer as the first frosts iced the muskeg moss, Ootek, Mowat's occasional Inuit companion, drew his attention to a barely audible sound on the wind.

'The caribou are coming,' he whispered. He had heard the faint cries of a wolf from an adjoining territory announcing the returning migration of the caribou from the north. Ootek gathered up his hunting gear and set off to intercept the herd, returning a few days later loaded down with fresh meat. Ootek related another story from his childhood on the tundra. Following Inuit tradition, when he was five years old his father, a powerful shaman, took the boy to a nearby wolf den and left him there. For twenty-four hours the young Ootek was fed by the wolves and played with the cubs, supervised all the while by one of the adults. Was this a modern example of the same sentiment behind the origin of the twin tracks of adult wolf and human child laid down on the floor of Chauvet cave some 30,000 years earlier?

I could have chosen any number of reports of humans and wolves living in close harmony in the wild. While none of these extends to witnessing the mutual cooperation on a hunt as I have imagined from the European Palaeolithic, the familiarity that the Inuit have with the wolves with whom they share the Barren Lands of Manitoba is a far better guide to understanding how we first grew close to the wolf than any amount of malicious falsehoods about their homicidal savagery.

Eventually Mowat's sojourn with the wolves came to an end and he returned home. He had found no evidence that the wolves were solely responsible for the fall in caribou numbers. It never made any sense to imagine that they were. After all, wolves had hunted caribou for tens of thousands of years before humans arrived in North America. Far from providing the excuse that his employers and the powerful hunting lobby were searching for to justify the final extermination of the wolf from the whole of North America, his report rather proved the complete opposite. It was filed under 'For Active Consideration' and never seen again.

10 Friend or Foe?

In light of these and many other very positive first-hand accounts of mutual interaction between wolves and men, how can we explain the fear and loathing that have been so prevalent in our recent history? Indigenous peoples like the Inuit and many Native American tribes certainly do not share it. They have retained an intimate relationship with Nature and often feel that the spirits of man and wolf are one and the same. For most of us who have lost touch with the wild world, the wolf became our enemy as soon as our ancestors abandoned hunting and took to raising animals for food. This has been a long process that began in the Middle East some 10,000 years ago. A conflict of interest between man and wolf began the moment that humans plucked animals from the wild and forced them into domesticated servitude. To hungry wolves, following their natural instincts and unrestrained by their instinctual taboo on attacking humans, domesticated animals were easy prey.

For millennia wolf and man had shared the wild country, each respectful of the other's boundaries. They had learned to live together harmoniously, so much so that they shared a common spiritual existence in a way that is familiar to Ootek, Mowat's Inuit companion, and to many other indigenous people. They would not have dreamed of killing a wolf. For them the wolf was an integral part of the land they both shared.

Over time in the 'civilised' world wolves have been punished for being themselves. They have been persecuted for centuries and hunted down all over the world. Where once they were found everywhere in North America, Europe and in Asia, the campaigns of extermination have eliminated the wolf from most of the United States, Mexico, Europe and the Far East. We have justified this brutal campaign by vilifying the wolf in many of our favourite stories.

I freely admit that I was not immune from the malign influence of these stories, as I discovered when I went to visit Shaun Ellis and his pack of captive wolves. On the train to Cornwall my pulse rate increased, I began to break out into a cold sweat at the mere prospect of coming face to face with a wolf. In the event I need not have worried. I was mesmerised. Their long legs, enormous paws, elastic gait and those intelligent, penetrating amber eyes instantly banished all fear and replaced it with a mixture of curiosity and admiration. Though I cannot pretend that this experience had the transforming force of Shaun Ellis's first encounter with the amber-eyed stare of a wolf in Thetford, I nonetheless returned from Cornwall with a completely different view of this beautiful species. I may not have much liked dogs, but I was captivated by the wolf.

By no means is everyone as convinced as Shaun Ellis or Farley Mowat that we are, or were once, capable of enjoying a close and cooperative working relationship with wolves. Quite the reverse. Brian Hare, the animal behaviourist, whom we shall meet later on, points out that throughout history no other animal has been portrayed so consistently as the Bad Guy. Children are taught to fear wolves with such tales as *Little Red Riding Hood* and *The Three Little Pigs*. From historical times to the present, wolves have been persecuted, hunted down and killed until now they are listed as endangered species in some regions. There are occasional redeeming stories such as that of the she-wolf that suckled Romulus and Remus, founders of Rome, after they had been left to die; but, by and large, we have viewed wolves with a mixture of fear and loathing. This makes it all the more remarkable that today we have chosen to take their closest cousins into our homes and into our hearts.

Shaun Ellis and others who have spent time with wolves are motivated to dispel their evil reputation, which they believe to be thoroughly undeserved. It is true that wolves do occasionally kill humans, but this is a vanishingly rare event. Humans, on the other hand, kill thousands of wolves on a regular basis. If our ancestors were once their close allies, what could have happened to transform them in our minds into the cunning evil creatures that nowadays we believe them to be? I am not alone in linking many features of the modern world – states, cities, complex and largely dysfunctional societies, even infectious disease – to the consequences of agriculture. Before that revolution, which began some 10,000 years ago, we lived as hunter-gatherers and left traces of our lives in places like Chauvet and Altamira. Although we had become the top predators of our time, we still depended on a deep understanding of the land and the animals with which we shared it, including, even perhaps especially, the wolf. When we left the life of the hunter-gatherer to raise crops and animals we were able to settle in one place rather than follow the herds on their annual cycles of migration. We no longer needed the wolf and we abandoned it.

In *Never Cry Wolf*, Farley Mowat puts the transformation of the wolf from friend to enemy like this:

> There is extensive evidence to show that far from being at enmity the wolf and hunting man worldwide enjoyed something approaching symbiosis, whereby the existence of each benefitted the existence of the other. But as European and Asian men began divesting themselves of their hunting heritage in order to become farmers or herdsmen, they lost this ancient empathy with the wolf and became its inveterate enemy. So-called civilized man eventually succeeded in

> totally extirpating the *real* wolf from his collective mind and substituting for it a contrived image, replete with evil aspects that generated almost pathological fear and hatred.
>
> European men brought this mind-set to the Americas, spurred on by bounties and rewards and armed with rifles.[1]

We turned the wolf from our friend into our enemy when its natural instincts led it to attack domestic flocks on which our own livelihood came increasingly to depend. We can get a sense of this whenever reintroduction is proposed.

When thirty-one wolves were moved from Alberta to Yellowstone National Park in 1995 after an absence of seventy years and in the face of energetic local opposition, they had an immediate and very positive impact on the park. Numbers of the wolves' main prey, elk and deer, fell rapidly and the damage these cervids had been doing by browsing on young trees was quickly reversed. But as the years passed, the influence of the introductions spread to other animals, not just their prey, and then to the physical landscape of the park itself. Numbers of elk and deer declined, but that was only the beginning. The deer learned to avoid parts of the park where they could be more easily hunted down. The containment of the cervids triggered an astonishing regeneration of trees, particularly along river valleys and in the many gorges. Before very long, the once bare hillsides were covered with quivering groves of aspen, willow and cottonwood. Songbirds returned. So too did beavers, which, by damming rivers, created shallow ponds for fish and other aquatic life. Wolves killed coyotes, which led to an increase in small mammals like mice and rabbits that in turn encouraged more hawks, bald eagles and other raptors. Bears benefited from the increased berry crop on the regenerating shrubs and their numbers began to rise. All of these welcome changes followed from the reintroduction of the wolf and spread down through what ecologists call a 'trophic cascade'.

There is a lively debate in Scotland at the moment. Using Yellowstone as an example, supporters of the wolf programme point out the advantages in cutting red deer numbers by reintroducing their one-time nemesis. But opponents raise the prospect of wolf attacks on humans and particularly on domestic animals like sheep and cattle. These familiar concerns are voiced around Yellowstone since wolves spread out from the National Park into the surrounding farmland as the numbers within the park increased. Wolves have been taken off the register of endangered species and hunting is once more allowed. It is proving difficult to find a balance between conservation efforts, aimed at maintaining a sustainable population of wolves in Wyoming and the neighbouring states of Montana and Idaho, and the hostility of a public brought up to fear the wolf.

In the version of events I am proposing, Palaeolithic hunters cooperated with wolves in killing animals. The wolves, lacking the equipment easily to dispatch a large animal such as a bison, risked injury when they killed slowly through blood loss. The wolves rounded up the prey, and held it at bay until the humans arrived to kill it with a spear. This instinct for cooperative hunting survives to this day. For example, a Rhodesian ridgeback will corner a lion until a hunter arrives to shoot it.

The kill was divided, the wolves taking the innards and the offal, the humans preferring the meat. The cooperation was only possible because of the similarities in social organisation which allowed wolf and man to work together, and this empathy was incorporated, over time, into the subconscious mind of both wolf and man. It was lost after the dawn of pastoral agriculture, starting about 10,000 years ago in the Middle East, when the wolf was painted as the enemy of man. However, the valuable atavistic empathy, embedded in the psyche of both species, was transferred by humans from wolf to dog. For the wolf's part, even though they learned to fear us, they never saw us as their enemy and still retain the ancient empathy. And that is why they did not eat Shaun Ellis.

Mowat's *Never Cry Wolf* was published in 1963 and went on to become a bestseller, eventually being translated into 37 languages. The Disney film of 1983 helped to spread the central message of the book: that, far from being murderous savages, wolves were gentle, family-loving, free-spirited creatures, misunderstood and wrongly persecuted. In the *Los Angeles Times Book Review*, David Graber wrote, 'By writing *Never Cry Wolf*, Mowat almost single-handedly reversed the public's image of the wolf from feared vermin to romantic symbol of the wilderness.' The villain of the piece, the Canadian Wildlife Service, which according to Mowat had bowed to pressure from the hunting lobby in an effort to exterminate the wolf from North America when they sent him up on a phoney mission, was deluged by complaints from outraged citizens. They responded by claiming that Mowat's account of his mission was factually inaccurate. For instance, he was not alone but accompanied at all times by two other biologists, and it had never been the intention of the Wildlife Service to exterminate the wolf. Nevertheless, however fanciful it may be, *Never Cry Wolf* is credited with starting to reverse the public's negative image of the wolf and setting in train the series of conservation measures, like the Yellowstone reintroductions, that continue to this day.

11 A Touch of Evil

There is always tension between accounts written by scientists, who in their professional work are taught to lay aside personal feelings and concentrate on the facts, and the wider public, who want to read a good story and do not share this obsession with factual accuracy at all costs. I once heard a very eminent embryologist, an author himself, commenting on the fussy detail of yet another fact-filled presentation at a conference. I don't remember anything about the paper but I do recall what he said when as session chairman he summed up the lecture: 'The trouble with you people [we were not embryologists] is that you know a great deal but you explain absolutely nothing.' To a greater or lesser extent, all authors necessarily struggle with the same conflict when they write for the general reader. *Never Cry Wolf* may at times have been fanciful, but millions have read it and been affected by it in a very positive way.

Fortunately, to balance this, there has been a good deal of scientifically rigorous field research into wolves living in the wild: perhaps not as entertaining nor as fanciful as *Never Cry Wolf*, but a source of factual information on this fascinating animal. Most of the field research has been done in North America, in Canada and Alaska in particular, where combined wolf numbers are estimated at around 20,000. They have been eliminated from the lower 48 states save for the reintroductions in Idaho and Wyoming and a small resident population in Minnesota and on Isle Royale in Lake Superior. Small numbers also live close to the Canadian border in northern Michigan and Montana and in the swamp woods of south-east Texas and Louisiana. Otherwise there are none. In Europe, where once they roamed freely, wolves have been entirely eliminated from the British Isles, where the last one was shot in Perthshire in 1680 by Sir Ewen Cameron. The last Scandinavian wolf was killed in Finland in 1911, but small numbers still remain in Italy, Spain and France. Further east, wolves are scattered in the forests of Estonia, Poland and the Balkans. They are wanderers and no respecters of political boundaries, so wolves can and do turn up almost anywhere on mainland Europe. I remember being in the South of France a few years ago when a lone wolf caused a stir by walking, at night only, along the main road from the Italian border. The local papers concocted a romance about a forlorn widower searching in vain for his lost lover. Outside Europe, wolves are found right across Asia, in Iran, northern India and Afghanistan and east into China, though their numbers are unknown. There is a rare wolf in Ethiopia, *Canis simensis*, but it belongs to a different species from the grey wolf *Canis lupus*.

In North America, the extermination of the wolf began in earnest when trappers moved west shortly after the Lewis and Clark expedition of 1804–06. Their expedition, commissioned by Thomas Jefferson, was sent to explore the lands west of the Mississippi that had been bought from the French in the Louisiana Purchase of 1803. The trappers were after beaver pelts and killed the wolves that had acquired the habit of stealing beavers from their trap-lines. By 1850 beavers had become so scarce that the trappers turned to killing the wolves for their pelts. At the same time buffalo came under intense hunting pressure, and by 1880 an astonishing 75 million of them had been slaughtered, mainly for their hides. Wolves learned to follow the buffalo hunters and feed on the carcasses left behind. They too came under attack from hunters, who rarely bothered to skin them. When all the buffalo had gone, men turned once again to wolfing for a living, and by the 1890s they had killed almost all the wolves east of the Rockies, from Texas in the south to Dakota hard up against the border with Canada in the north.

The Wolf Wars, as they came to be known, were accompanied by a campaign of vilification reminiscent of the vitriolic propaganda attacks which we have always used against our enemies, human or otherwise.

The extent of this slaughter, which sickens most of us now, astonished the Native American tribes who had lived side by side with the wolf for millennia. They had, and still have, a completely different attitude to the land and the animals with whom they share it. If you will accept the generalisation, Native Americans see themselves as the custodians of the land and the animals under the watchful eye of the Great Spirit. Of course, they still hunted buffalo, elk and deer for food and clothing, but not before offering prayers that asked for the animal's blessing. And herein, in my own view, lies the answer to the whole question of our relationship with, first, the wolf, and now the dog. Unlike us 'civilised' Westerners, Native Americans understand that some things are inexplicable except through the medium of myth.

We may scoff at these 'primitive' and unscientific explanations for natural phenomena, but they closely resemble the way our own Palaeolithic ancestors saw the natural world, as is evident from the painted galleries of Chauvet. There are innumerable Native American legends involving the wolf. Barry Lopez, in his 1978 book *Of Wolves and Men*[1] recounts one of them from the Cheyenne. In common with other Plains tribes, the Cheyenne created societies of young warriors whose task it was to defend the tribe against attack and also to lead raiding parties against their neighbours. The 'Wolf Soldiers', as they were called, rose to prominence in the early nineteenth century as the Native American Wars began in earnest. At the time, the two main branches of the Cheyenne were each travelling northwards towards their traditional lands in the Dakotas from which they had been forcibly deported to reservations in Oklahoma. The leader of the Wolf Soldiers, Owl Friend, set out alone one night to join up with the southern Cheyenne when he was caught in a sudden snowstorm. In the darkness he came across a single lodge pitched by the side of a stream. He approached the entrance, where he was met by a group of young men who welcomed him inside. They gave him food, started to dry his clothes that had been soaked by the storm and put him to bed. The storm lasted for four days. On the morning of the fifth day the young men led Owl Friend outside. The storm had abated and the sky had cleared. 'Remember this,' they said, 'We give it all to you.'

The next morning Owl Friend awoke to find himself in the middle of the open prairie. He was surrounded by four wolves, whom he recognised as the young men from the lodge. 'Repeat this dance for four days and four nights and you will become a Wolf Soldier,' they said. Owl Friend did as they said. On the fifth day he returned to his tribe and inaugurated the Wolf Soldiers of the Cheyenne, the last and the most feared of the seven great Cheyenne soldier societies. In the years following their inauguration the Wolf Soldiers fought their enemies with great courage and ferocity, qualities that, through the power of Owl Friend's dream, had come to them directly from the wolves.

Similar myths of spiritual transference crop up again and again in Native American legend. All we have in Europe to remind us of the ancient spiritual affinity of man and animal are the murals of Chauvet, despite the strange fact that none depict wolves. But there is a much more vivid reminder even than art or myth of the ancient pact between wolf and man – and it lives on in the minds and the lives of dog-lovers everywhere.

The wolf has been demonised throughout history as the embodiment of evil, savagery, even wanton lust and sexual promiscuity. In medieval Europe, people accused of being werewolves were allegedly capable of alternating between human and lupine forms. They were burned at the stake. We now believe these unfortunates were nothing of the kind but likely to have been suffering from a severe form of schizophrenia. Feral children raised by wolves, which has happened occasionally, often exhibit symptoms interpreted as severe autism when 'rescued'. They huddle in a corner, never speak, refuse clothes and are prone to outbursts of unprovoked aggression. In neither case do we now believe that the people have literally become part-wolf, part-human.

Mythological wolves were rarely pictured as anything other than the epitome of evil. The Norse myths tell of the sinister wolves Fenris and Garm, and the shape-shifting Loki, held in chains until the end of the world, the Twilight of the Gods, when they broke free to fight the Norse god Odin and the Aisir to the death at Asgard. The wolf's association with death and destruction is everywhere. The Church aligned the wolf with the Devil, and during the reign of terror implemented by the Spanish Inquisition beginning in 1478, his agents, the unfortunate werewolves, were hunted down and put to death.

This deeply embedded hatred of the wolf makes it all the more surprising that dogs, which are so little removed from wolves, have become 'our best friends' and the source of so much devotion, happiness and, yes, even love. Can it be that we recognise so much of ourselves in the dog and indeed the wolf that our emotions are amplified? Is what we see when we look into the eyes of either animal a reflection, a transference of our own emotions? Shaun Ellis certainly experienced this when he first met the wolf's amber gaze. He felt its eyes drilling into his very soul.

He later reflected that perhaps he was completely mistaken and that the wolf felt nothing of the sort and was probably just thinking about its next meal. Dog owners consistently describe the same feeling of mutual understanding as Shaun experienced, but is it real? And does it really matter anyway? After all, the most obliging server in a restaurant may think all her customers are complete fools, but as long as she doesn't spill the soup, however tempting that might be, they are none the wiser. Most dog owners would fiercely contradict this interpretation of their pet's behaviour towards them, and they may well be right. Perhaps the admiration they see in their pets is real and well deserved. But as with the restaurant server, does it really matter if it is a pretence, so long as they manage to keep it up?

12 The Basic Framework

The mesmerising stare of the wolf which many of us, including myself, find so bewitching may have an entirely different purpose for the wolf. Before a pack chooses which animal in a herd to pursue, wolves look closely at each animal in turn, and more often than not the animal returns the gaze. Native Americans call this the 'Conversation of Death', and its purpose is for the wolf to assess the condition of the prey. The healthy animals in a herd often completely ignore a pack of wolves as they know they will pass this inspection. The sick and ill betray their nervousness to the wolves by their actions, sometimes even going so far as to stand up and walk away from the rest of the herd as if giving themselves up. The accuracy of this assessment is of paramount importance to the wolves, as the correct choice of victim may mean the difference between a full stomach and starvation. Bear that in mind the next time you exchange fond glances with your pet.

I was told as a child that dogs could smell your fear, so never show it. Of course, that made things even worse and the Hell Hound down the road must have realised I would make an easy kill, which no one will persuade me was not his intention every time I passed his gate on the way to school. Dogs are not wolves, but although many of their behaviours have been sublimated over the millennia as we have grown closer, the original template remains firmly lupine.

Whether or not we like to think so, everything about a dog – its appearance, its senses and its behaviour – are just versions of the wolf's, albeit attenuated or enhanced by millennia of selection. Selection has done amazing things to create today's dog, but the scaffolding on which it is built remains that of the wolf, and the limits to what selection can achieve are set by this basic framework. Selection cannot produce dogs that naturally walk on two legs. Although this is a trick that can be taught, no dog will ever walk on two legs better than it does on four, at least not in the foreseeable future. They are constrained by the architecture of the hip as surely as our own ancestors were in the distant past. Likewise, a dog will never learn to talk, though again they are able with training to appear to mutter a few words. The essential neurological pathways required for language are simply not there. Nonetheless, within these limits our two species have discovered similarities that have brought us together despite the vast evolutionary distances between us and the enormous differences in our genomes.

We can form much closer emotional bonds with dogs than we can with our nearest evolutionary cousin, the chimpanzee. Experts in animal behaviour have been aware of this for years and tested it thoroughly. For example, even though chimpanzees are much better than dogs at performing a wide range of tasks, dogs do better than chimps in others. One popular test is to place an item of food under one of two upturned beakers then see how often the dog or the chimp picks the correct one at a given signal from the experimenter. Given a dog's extraordinary sense of smell, the experiment incorporates controls to avoid the animal using this advantage, for example by smearing both beakers with food or by hiding them behind a glass screen. With no other clues, both dogs and chimps select the correct beaker about half the time, as expected. The experimenter then does something to indicate which of the beakers covers the food, perhaps with a glance or by touching the beaker, and then counts the successful attempts. Chimps are not good at this test, although they will eventually learn. Dogs, on the other hand, pass with flying colours almost immediately.

The conclusion of this and other tests is that, while the chimps can eventually work out what to do, the dogs are able to follow the actions and read the intentions of the experimenter straight away. The interpretation of this uncanny ability, which is also shared by the wolf, is that it stems from their acute sensitivity to the signals from other members of the pack. They are also able to read the signals given out by prey animals that help the wolves to select the right target, catch it and kill it. On the open tundra, knowing which of a hundred caribou to go after is vital to avoid a fruitless pursuit. This is the ultimate origin of the dog's ability to pick up the subtle signals that their owner is about to take them for a walk well before the door is opened. The natural inclination to interpret this as a sign of intelligence in our beloved pet is countered by their stubborn inability to perform tasks that we see as very straightforward. For example, faced with food placed behind a length of fencing, a dog will repeatedly try to get through it, ignoring the encouragement of the owner to go round instead.

It's a long time since our ancestors survived on hunting wild game, and since the invention of agriculture we have almost entirely lost the ability to read the minds of other animals. Nonetheless, many hunters report a feeling that they 'know' what their prey will do next, another of the many atavistic senses from our Palaeolithic past. We were hunters once, and in many ways, deep inside, we still are.

There is one other vital feature that we share with the wolf that has not yet been completely extinguished by millennia of civilisation. Like the wolf pack, we depend on each other. Surprising as it may seem, and however imperfectly we manage to do so individually, our success as a species stems from our ability to work together, just like the wolf pack, to achieve what we cannot do alone. The basic unit of cooperation in humans has always been the family and there are good reasons for this.

When we slowly evolved from tree-living ancestors in the African forests and became truly bipedal as we colonised the savannah, our skeletons had to change to allow us to walk upright. One of these changes was in the pelvis, where the pubic bone fused at the front, greatly narrowing the birth canal. Simultaneously and in order to grow a larger and more complex brain, babies' skulls grew and, ever since then, labour and birth have been difficult and dangerous for women. The growth of the brain and the time taken to make all the right neuronal connections within it extended the time that babies and infants need our care and attention. This meant that to survive during the long time it took for our children to become independent, mothers needed the support of the family group. Also we learned the advantage of hunting in groups rather than alone. That cooperation enabled our ancestors to hunt much larger prey, and this was particularly important in Palaeolithic Europe where, as we have seen, there was an abundant supply of large herbivores.

Even though our distant African ancestors probably encountered Cape hunting dogs and allied species that also hunt in well-organised packs, when they first arrived in Europe some 40,000 to 50,000 years ago they would have seen at first hand the hunting strategy of the grey wolf and perhaps even begun to emulate it. Although humans and wolves had little else in common, the ability to cooperate within a close-knit family group was very strong in both species.

The basic social unit is the wolf pack. It is almost an organism in its own right, separate from the individual members and whose survival as a unit is paramount. A wolf that finds itself excluded, excommunicated from the pack, must find another one and be accepted into it or face starvation and an early death. A typical pack is an extended family of around half a dozen related individuals. Occasionally packs of thirty wolves are reliably reported, but hundreds-strong 'super packs' gleefully reported in the press make for good copy but are highly unlikely. In a typical example,

'According to reports ... a state of emergency was declared when the eastern Siberian town of Verkhoyansk on the banks of the Yana river was surrounded by four hundred ravenous wolves who killed thirty of the village's horses, leaving the terrified villagers wondering whether they would be next on the menu.'[1] This story plays to our deep-seated fear of wolves, which, together with our fear of the dark, deep forests and dangerous predators, is another aspect of the Palaeolithic archetype which is part of our make-up even today.

Life in the wolf pack is much more prosaic. As we have seen, there is usually just one breeding pair in each pack, the alpha pair, and breeding takes place only once a year, in February or March, with cubs being born after sixty-three days. Litters are usually between four and six, but the number varies depending on the availability of food. In good years there might be up to a dozen pups in a litter, while in lean years the alpha pair may not breed at all. Pups are born deaf and blind. They can hear after a few days, open their eyes about ten days later and are weaned at five weeks. By this time they have ventured outside the den, but they stay close to the entrance while they play. The play begins to establish the all-important social order within the litter that, though it changes many times, underlies the ultimate organisation of the pack. Dog owners and breeders will recognise many of the same milestones.

While we see ample demonstrations of love and devotion in a mother's treatment of her cubs, it is underscored by the ruthless discipline of the wild. Any cub showing signs of physical disability will be killed, and even eaten. Each of these measures, the hormonal regulation of litter size according to the abundance of prey and the elimination of the weak and disabled, is there to protect the survival of the pack. So too is the interest the other adults show in the alpha female's cubs and the affection the youngsters give in return. Unlike in many other species, adults make no attempt to take food given to growing pups, and within the pack there is a strong sense of respect for the social order.

This is reminiscent of the way many social insects like bees and ants behave where the lives of the workers, all sterile, are completely devoted to the survival of the alpha female's, the queen's, offspring. The evolutionary explanation for this altruism is that even though the workers have no offspring of their own, their efforts to support their sister, the queen, help to ensure that their DNA flows through to the next generation. As with the beehive, the wolf pack is at heart an extended family. Konrad Lorenz, a keen observer of dog behaviour, interpreted the altruism of the wolf pack as marking the beginning of a primitive sense of morality. The strong sense of family shown by the wolf, and shared albeit clumsily by ourselves, lies at the heart of the unexpectedly close relationship our ancestors managed to develop with this wild creature and which some of us continue to enjoy with our pets.

A grey wolf pack at rest in a wolf reserve in Canada. The social bonds of the pack are key to the wolves' happiness and survival.

Most research on wolf behaviour, especially in the early days, was carried out on captive animals confined to an area only a fraction the size of their natural territory. These animals were fed every day, so did not need to hunt, neither were they necessarily related to each other, as is the case in the wild. John Bradshaw from the University of Bristol specialises in the behaviour of dogs. He has come to think that the commonly held view of a rigidly enforced dominance hierarchy within wolf packs is seriously distorted simply because the animals observed in captivity are unrelated.[2] Captive animals tend to be more aggressive towards each other than is generally seen in the wild, and this is no surprise in Bradshaw's view. Basing theories about pack behaviour on captive wolves would be like inferring all the subtleties of normal human interactions from the behaviour of long-term prison inmates.

According to Bradshaw, this misinterpretation of pack behaviour, in particular the role of the 'alpha male', is responsible for some of the more distasteful methods of dog training, often involving physical punishment. Adolph Murie, the first biologist properly to study wolves in their natural Arctic habitat, was left with the impression that first and foremost wolves within a pack were friendly to one another.[3] Unlike captive wolves, they absolutely have to get along to survive. Although there is a harsh side to discipline, it does not detract from the amicable and mutually supportive relations that are the fundamental adhesives of the pack.

Although the structure within a pack is fluid and can change with time and circumstance, there are two separate hierarchies. The alpha female and the alpha male take precedence over the others of their sex, and although it was once thought that they formed the only breeding pair, it now emerges that although the alpha female is the only animal to have young, it is not always the alpha male that is the father. Again, rather loosely, precedence is signalled by a variety of postures and gestures, and generally, but not always, the alpha animals are the first to feed at a kill.

The myth of the alpha male who must never be challenged has become an easily recognised caricature of our human society but is based, at least in part, on our distorted impression of the social order of the wolf pack. It is the alpha female that has the greatest say. She selects the victim and directs the hunt. The beta animals are the ones that enforce the will of the alpha female, for example by challenging the prey to help her gauge its fitness. Shortly before the cubs are born, the pack converges near the denning site chosen by the pregnant female, to help out. They bring her food and, after the pups are weaned, they take over responsibility for teaching them the ways of the pack.

The parallels with our own family lives are striking, and therein lies the root of the bond between the two species. The wolf has by necessity honed its social arrangements to fit its method of cooperative hunting, and this spills over into their family life in ways we find very familiar. When our Palaeolithic ancestors first encountered the grey wolf in Europe, they saw a wild animal that behaved on the hunt much as they did themselves. Their predecessors, the Neanderthals, must also have witnessed wolves in action, but as far as we know never contemplated working together with them for mutual advantage. Textbooks speak of the domestication of the wolf as if it were an inevitable process directed solely by and for the benefit of humans. I prefer to think that it was not inevitable at all but required a definite inventive step by one or a very few individuals, and we will examine the genetic evidence for this statement a little further on in the book.

Even these days very few original inventions are made by more than one person. After the all-important inventive step there may be many modifications and refinements, but truly original thoughts are very few and far between. Versions of the invention, if it is any good, spread quickly – these days almost instantly – but the original thought happens only once. I can think of several cases in science where this has happened. The discovery of penicillin by Alexander Fleming, the revelation of the basis for genetic fingerprinting by Alec Jeffreys, the polymerase chain reaction which transformed molecular genetics by Kari Mullis, and the discovery of the helical structure of DNA itself by James Watson and Francis Crick – all of these sprung from one or two brilliant minds. The same is true in all branches of science, technology and the creative arts.

The alliance our ancestors forged with the wolf was an event of seminal importance in human evolution and ranks with the other innovations such as the bow and arrow, the spear-thrower and the flowering of art and music. All these things secured our ancestors' survival through the harshest conditions of the Ice Age.

13 We See the First Dogs

We must wait until about 12,000 years ago before we get the first unmistakable evidence of the intimate bond developing between man and dog. This came from Natufian burials at Ain Mallaha, in Israel. The human inhabitants, the Natufians, were Mesolithic hunter-gatherers, but, unlike their ancestors in the Upper Palaeolithic whose nomadic life was dominated by the migrations of prey animals, they led a more sedentary life in rich surroundings. The region was heavily wooded, with abundant sources of nuts and wild grains, and fish in nearby Lake Hula. Though they were not yet truly farmers in the modern sense they were certainly heading in that direction. Rather than moving from one temporary camp to another they built proper houses – large circular stone huts, set into the ground, with roofs supported by wooden poles.

The Natufians buried their dead in their own houses, which they afterwards abandoned. Central to our story is one particular burial of an elderly woman, her head resting on the body of a small puppy. This is the first proof of what the authors of the paper reporting the excavation cautiously refer to as 'an affectionate rather than gastronomic relationship' between the two burials.[1] They must have been very close indeed.

By this time the agricultural revolution was well and truly under way. Arable crops such as wheat and barley replaced the harvest of wild plants.

Domesticated animals, which could be easily confined, made meat supplies more reliable. No longer constantly on the move, the people followed the Natufian example and settled down in one place. Feeding themselves and their families no longer took up all their time, and culture, in the form of music and art, flourished. Alongside these welcome improvements to daily life came the emergence of new ideas. Concepts of property and ownership, unknown to our free-living ancestors, started to divide society between the haves and have-nots. The egalitarian social structures of the Palaeolithic hunter-gatherers diminished, labour was for hire, and before very long, servitude and slavery took their place. The invention of agriculture also heralded the abrupt shift in the human attitude to the wolf and ended the 30,000 years of mutual cooperation that had kept our ancestors alive through the hard times of the Upper Palaeolithic.

The first clear depictions of a dog were discovered in Mesopotamia. One is a small gold pendant found in the ancient Sumerian city of Uruk on the banks of the Euphrates River in southern Iraq. Uruk was the capital of Gilgamesh, and it had embraced the trappings of urban life: a military garrison and a full-time bureaucracy overseeing a many-layered society. The pendant, dating from about 5000 years BP and currently on display in the Metropolitan Museum of Art in New York, is clearly modelled on a dog, and not a wolf. It is striking in that this creature had already acquired one of the features often associated with domestication, a tail curling over its back, which is reminiscent of a modern Husky, Samoyed and other spitz breeds.

It is not obvious exactly what the Uruk dog might have been used for – guarding or possibly just companionship. Later, stylised images of dogs began to appear on ceramic shards. In two small farming villages in Iran, Tepe Sabz and Chogha Mish, the dogs looked very like the ubiquitous guard dogs still plentiful in the region today. Again, their tails are curled over their backs.

Another indication of what dogs were becoming came from the large prehistoric settlement in Susa, in western Iran, dating from around 6000 BP. Over 1,000 painted vessels, mainly pots, beakers and jars, were recovered from a large cemetery and were clearly intended to accompany the dead into the afterlife. Many displayed stylised images of birds, insects, reptiles, and occasionally of dogs. The dogs closely resembled modern salukis with their lithe bodies, deep chests and long legs. These dogs were almost certainly used for hunting, just as in modern times. Until the last few decades, salukis were still used to bring down gazelle, fox and ibex and, even today, are worked with falcons. Given the similarity of the Susa images to the modern saluki, it looks as though very little has happened to change their overall appearance over the last 6,000 years. Their mythical associations first appear in *The Epic of Gilgamesh*, dating from around 4000 BP. This text describes in words, though without illustration, the goddess Innana travelling alone with seven hunting dogs in tow, each with a collar and on a leash.

In contrast to the depictions of hunting dogs built for speed, clay figurines from about 4000 BP have been excavated at Nineveh in northern Iraq close to the modern city of Mosul. These figurines show large dogs resembling modern mastiffs, whose principal purpose is made clear by the inscription on one of them: 'Don't stop to think. Just bite.'

Dogs maintained close emotional ties with their human masters, as is clear from their inclusion in human burials, as at the Natufian site of Ain Mallaha. By the fourth century BCE, much more defined images appeared on two 'palettes' excavated from an archaeological site associated with the Naqada culture, which flourished in pre-dynastic Egypt between 3500 and 3200 BCE. These palettes, used to grind the ingredients of an early form of mascara, were often richly decorated. As an example, the so-called Hunting Palette depicts a lion hunt, with five arrows bristling from the beast's neck. Clearly visible are three dogs, not wolves by any means, but of a more slender build similar to the Mesopotamian examples.

In ancient Egypt the dogs themselves were venerated by ceremonial burials as they passed into the afterlife. A celebrated example comes from the Old Kingdom site near Giza around 4500 BP where a named dog, Abuwtiyuw, belonging to an unknown servant of an unknown king, was buried beneath a limestone slab bearing the following inscription: 'His Majesty ordered that he [i.e. the dog] be buried ceremonially, that he be given a coffin from the royal treasury, with linen and incense.'

This was clearly a very special dog.

Dogs were highly regarded in ancient Egypt and, when families could afford to do so, they had their dog mummified and ceremonially interred amid great displays of sorrow, even extending to grieving family members shaving off their eyebrows.

Abundant Egyptian tomb paintings give us a good idea of what dogs looked like in later dynasties. For instance, the magnificent tomb of Ramesses II in the Valley of the Kings, dating from 3213 BP, is richly decorated with bas-relief panels. One shows the pharaoh with a team of slender saluki-like hunting dogs, ideal as sight-hounds in the open desert terrain where herds of gazelle and oryx were abundant.

The slightly earlier tomb of Tutankhamen contains probably the most familiar image of a dog from the ancient world – a magnificent black statuette of Anubis. As a god, Anubis is said to have the head of a jackal, but his statue in Tutankhamen's tomb is all dog by the look of it. With its large erect ears and slender body, the statue closely resembles the earlier Mesopotamian images of sleek hunting dogs.

Later in history dogs are depicted in many tomb paintings, sometimes with a strikingly piebald appearance reminiscent of today's Dalmatians. Mummified dogs were frequently buried with their owners. In the catacombs at Saqqara, close to the temple of Anubis, the mummified remains of a staggering 8 million dogs sacrificed to the god were discovered in 2015.

This Anubis shrine, made of wood, plaster, lacquer and gold leaf, was found in the eighteenth-dynasty tomb of the pharaoh Tutankhamen.

The fondness of the Egyptians for their dogs is evident from the names they gave them, preserved on their collars: 'Brave One', 'Reliable', 'Good Huntsman' and even 'Useless'. This tells us that the qualities of loyalty that dog-owners prize most highly today were already well developed many thousands of years ago.

As well as hunting, the other main purpose of these early dogs was guarding, and for that task a more solid physique was developed, best seen today in mastiffs. We see this particularly in Roman wall paintings where images of dogs very similar to modern mastiffs are being used for guarding sheep flocks and protecting human settlements and, lastly, as weapons of war.

The preferred guarding breed, both in ancient Greece and in Roman times, was the Molossian. This was a huge dog whose origin has been attributed to the legacy of Alexander the Great's military expeditions to

India and Afghanistan. Allegedly, he was so impressed with the breed that he sent some specimens back to Greece, where they became the forerunner of the mastiff. The Molossian was also the favourite breed of the armies of both Greece and Rome. Although they were mainly used on patrols and as sentries, Molossians did fight in battles side by side with the Roman infantry.

Looking back at descriptions and representations of dogs in past times conveys the evolution of physical features over time. In Mesopotamia and in Egypt dogs had already undergone a dramatic change from their lupine ancestors in the Palaeolithic. Two quite separate types had emerged – the guard dog and the hunting dog. Though they could not yet be called breeds in the modern sense, we can assume that they evolved from a common stock by selection from the largest, fiercest, most obedient or swiftest adults, depending on the desired qualities. There was no need for breed isolation to maintain a standard. As we will go on to explore, the genetics of modern breeds reveals the fairly relaxed breeding regimes of earlier times.

We need no longer rely solely on ceramics or carved images to discover what dogs were like in ancient Rome. This is thanks to a twelve-volume treatise on farming, *De Re Rustica*, written by Columella around 60 CE and discovered intact in a monastic library in Switzerland during the fifteenth century. After a career in the army, Columella was a military tribune in the Roman province of Syria. Eventually he retired to his farm in Italy and immersed himself in its improvement. He emphasised that every farm needed a dog.

> Buying a dog should be among the first things a farmer does, because it is the guardian of the farm, its produce, the household and the cattle. It should be big and have a loud bark, first to intimidate the intruder when it is heard and then when it is seen. Colour too is important. An all-white dog is recommended for the

> shepherd to avoid mistaking it for a wolf in the half-light of dawn or dusk, and an all-black guard dog for the farm to terrify thieves in the daytime and be less visible to trespassers at night. It should not be too savage, so as not to attack the inhabitants of the house, nor so *mild* that it fawns over the thief. The farm-yard dog should be heavily built, with a large head, drooping ears, bright eyes, a broad and shaggy chest, wide shoulders, thick legs, and short tail. Because it is expected to stay close to the house and granary, its lack of speed is not important. The sheepdog, on the other hand, should be long and slim, strong and fast enough to repel a wolf or pursue one that has taken its prey.[2]

Being a farmer, Columella has most to say about guarding rather than hunting dogs. For information on the latter we turn to the earlier writings of the Athenian polymath Xenophon, a contemporary of Plato, writing around 370 BCE. Xenophon distinguishes two types of hunting dog, the Laconian and the Vertragus. The Laconian hunted by following a trail, like any scent-hound today. The Vertragus, on the other hand, chased down its prey by sight, and hunters needed to be on horseback to keep up with it.

Odysseus owned a Laconian, named Argos – meaning 'swift-footed' – and trained it from a puppy to hunt wild deer, hares and goats. Argos was his master's constant companion until Odysseus set out for the Trojan War. It would be another twenty years before they met again after Odysseus's eventful journey back to Ithaca, as described in Homer's epic poem *The Odyssey*. On his return Odysseus discovers his house is besieged by suitors all keen to impress and marry his wife Penelope. Planning to slaughter them all, he enters his house dressed as a beggar but discovers Argos on the threshold, old, weak and neglected. Even so, the dog recognises him but only has the strength to wag his tail before he collapses, dead, at Odysseus's feet.

The only other distinct type of dog known from the ancient world was much smaller than either mastiffs or hunting dogs and was used for an entirely different purpose. From the rare surviving images, notably on a Greek amphora from about 550 BCE found in the Etruscan city of Vulci, this small breed was similar in appearance to the Maltese, a modern toy breed. One of a clutch of modern alternative names for the breed is 'Roman Ladies' Dog', and the historian Strabo, writing in the first century CE, tells how it was favoured by wealthy patrician ladies. Was this the first record of the modern 'pampered pooch'?

By the classical period, dogs had evolved into clearly distinct types, not quite breeds in the modern sense, but evidence that enormous changes in appearance and behaviour compared to their lupine ancestors were well established over 2,000 years ago.

Though Columella the farmer offered plenty of advice on the right choice of dog for guarding the flock and the farmyard, he was silent about another feature of farm life, with which he would have been familiar – the presence of rats. All dogs are good at catching these destructive pests, which, like dogs, have discovered the benefits of living with humans but not the trick of making themselves appealing to their patrons. As far as I am aware, none of the ancient breeds was especially bred to catch rats, but their eagerness to do so would not have escaped notice. Besides being very useful in killing vermin, catching rats with dogs appears to be enormous fun.

While the aristocracy hunted game on their estates, the landless working man used dogs to kill rats. Modern terrier breeds all began as rat-catchers. The sport of rat-baiting became widespread in the eighteenth century, with enclosed pits used to contain competing terriers and an assortment of live rats. Onlookers placed wagers on their favourite dog and, if it had killed the most rats in a bout, collected the winnings. These assemblies had a historically significant role. They were the forerunners of the modern dog shows, as fashion, and the law, turned against cruel sports during the nineteenth century and terrier owners funnelled their enthusiasm into showing their dogs in competition instead.

14

The Studbook of Dudley Coutts Marjoribanks

At the end of the eighteenth century, while dog varieties ebbed and flowed in popularity there were no established breeds in the modern sense of the term. Among the British landed classes, dog varieties clearly similar to modern spaniels were bred to assist gentlemen in finding the birds that they had shot down. However, until the late eighteenth century, the slaughter of wild birds was limited by the technology of guns and particularly their rate of fire. After each shot the muzzle-loading guns of the period took a long time to reload and this meant that only a relatively few birds could be downed in a day's sport.

Although Henry VIII was said to have owned a breech-loader, it was not until 1772 that Patrick Ferguson, a British officer serving in the American War of Independence, invented the first practical breech-loading flintlock. The design took off and before long breech-loaders were the popular choice for shooting among gentlemen of leisure. The effect of the innovation was to greatly increase the kill rate of a shooting party. With that came the desire for a new type of dog capable of retrieving birds quickly, in large numbers and without damage.

Dudley Coutts Marjoribanks, son of a partner in the exclusive London bank of the same name and a keen sportsman, acquired sufficient personal wealth to purchase the 2,000-acre Guisachan Estate in Glen Affric, near Loch Ness in the Scottish Highlands. He served as Member of Parliament for Berwick-upon-Tweed before being created 1st Baron Tweedmouth. Like many of his Victorian contemporaries, Marjoribanks was an enthusiastic naturalist and in 1868, he set out to create a new breed of dog by scientific means that would combine all the features required for a modern gun-dog: speed, of course, an acute sense of both sight and smell, intelligence and, importantly, a soft mouth to carry the downed birds back to the guns.

Others in the past had bred for desirable traits, but what sets Marjoribanks apart is his meticulous record-keeping. I have marvelled at his studbook, preserved in the library of the Kennel Club's Mayfair headquarters. In the best copperplate script, he details all the breeding pairs at his Guisachan kennels. There is recorded the careful creation of the first of the nascent new breed: Cowslip the Golden Retriever from Nous, a yellow wavy-coated retriever, and Belle, a Tweed Water Spaniel. The family tree carries on with contributions from Sambo, a black Labrador and a sandy-haired bloodhound, down to Queenie, at which point the breed was established and closed to outsiders.

What this meant in effect was that Queenie's Golden Retriever descendants, and that means *all* pedigree Golden Retrievers, had managed to survive from that point onward with a severely restricted gene pool, limited to the genes inherited from Queenie and her male companions. It is appropriate that the Tweedmouth studbook is housed in the Kennel Club library because one of the important activities of the club, which continues to this day, is to maintain a register of pedigree breeds. The Kennel Club was founded in 1873 in Victoria, London, by 'ten gentlemen', and, although it has grown enormously in size and influence, it still maintains the atmosphere and traditions of a London gentlemen's club. I have visited the Kennel Club on several occasions in the course of my research for this book and have been warmly welcomed every time. Like other London clubs, the walls are hung with portraits, but other than Her Majesty the Queen, the Club's patron, all the others are of dogs.

We have touched on the Victorian working man's enthusiasm for competitive dog shows as an alternative to ratting contests. The first dog show was held in Newcastle upon Tyne town hall in 1859 and, for the country gentleman, the inaugural field trial took place in 1865 in Southall. The formation of the Kennel Club eight years later was initially intended to impose some standards on dog shows and field trials that would encourage fairness and the welfare of the animals taking part. The sport enjoyed an explosive growth, fuelled by the Victorians' enthusiasm for hobbies and natural science.

Darwin's *On the Origin of Species* was published in 1859, the same year as the first dog show was held. The book introduced the public to the concepts of evolution and selection which they could observe in their own dogs. In the wild, Darwin argued, the inherent variation existing between individuals was the bedrock of evolution. The better the individual fitted the environment, the more offspring it had and the greater the chance that the genes responsible for the advantage would be passed on and spread in subsequent generations. The 'environment' in this case was their natural surroundings, and so Darwin gave his book the full title *On the Origin of Species by Means of Natural Selection, or the Preservation of Favoured Races in the Struggle for Life*. The book captured the mood of the time and became an instant bestseller.

Darwin also recognised, however, that selection was not always 'natural'. It could also be 'artificial'. Taking the fantastic range of freakish pigeon varieties created by breeders as a working example, Darwin suggested these were mutants (he used the term 'sports' to describe them) which stood no chance of surviving in the wild and were only perpetuated by artificial means, in this case by the pigeon fanciers. The Victorian craze for novelty also caused dog breeders to pick out the rare 'sports' in their litters and breed from them.

When Lord Tweedmouth was perfecting the Golden Retriever he was selecting naturally occurring rather than freakish variation, but some breeds do have their origins in 'sports' that would certainly not survive for long in the wild. The truncated limbs of the Dachshund, the extravagant coat of the Komodoro, the compressed face of the French Bulldog, all came from 'sports'.

Darwin knew nothing of genes or DNA, whose discovery lay years into the future. Nevertheless, 'sports' and natural variation are both ultimately caused by genetic changes. There is no fundamental difference between the two. Only the means of propagation differentiates the 'sports' that need humans to help them survive, and naturally occurring variants that do not.

As we have seen, dogs began their evolution toward the extraordinary range of creatures that we see today a very long time ago. First to emerge were dogs bred for guarding or for hunting. These early specialists were certainly the result of selective breeding, although the strict rules of the pedigree dog were not enforced as they are today. Special traits were enhanced by breeding from dogs that displayed them. Breeds were not closed, but neither was breeding an entirely random affair. Thus, for centuries before breeds became officially recognised and standards were set, different breeds, or should I say 'proto-breeds', were recognised. Many 'proto-breeds' carried the same name as their modern-day pedigrees, though they did not conform to any standard.

It is clear from the archaeological and pictorial evidence from Mesopotamia, Egypt, Greece and Rome that the differentiation between dogs was already apparent in the early historical period. The two most obvious early specialists were large robust dogs, like the Molossian, used for guarding and war, and slim, saluki-like hunting dogs built for speed.

As time went on and dogs, through selective breeding, became more and more specialised for particular roles, a whole range of proto-breeds emerged. Sporting dogs were divided naturally into sight-hounds, like the Greyhound, which search for and pursue their prey using their eyes, and scent-hounds, like the bloodhound, which use their noses. Both these qualities, keen sight and a sensitive nose, are part of the sensory armoury of the wolf and clearly have a genetic basis, without which no amount of selective breeding could enhance these qualities. By breeding from the dogs with either the best sense of smell or the keenest eyesight the proto-scent-hound and proto-sight-hound breeds became differentiated. This

An Irish Wolfhound and a Toy Poodle–Chihuahua mix illustrate the remarkable range of physical variation within the species *Canis lupus familiaris*.

was Darwinian evolution in an artificial setting. The naturally occurring variation within dogs was channelled into creating what looked like a new species, but in fact was not. They were still members of the same species and, like all dogs, were still capable of interbreeding but were prevented from doing so by their owners.

15
The Emergence of Modern Breeds

By the middle of the nineteenth century, dog breeds became formalised, first by the Kennel Club and shortly after by its North American equivalent, the American Kennel Club (AKC), formed in 1887 from the amalgamation of US and Canadian breed clubs. As well as coordinating the increasingly popular dog shows and field trials, both clubs were concerned about maintaining breed purity. The AKC stated as much in the aims set out in its prospectus: 'to do everything to advance the study, breeding, exhibiting, running and maintenance of purity of thoroughbred dogs.' Both organisations have worked closely together from the outset and, as well as organising dog shows and a host of other activities, both maintain breed registries which are updated on a regular basis. New registrations are accepted only if both parents are of the same breed. This requirement brings its own complications, as we will see.

Both clubs publish registration statistics. The Kennel Club registers 250,000 new dogs every year and the AKC figure is approaching 1 million. At the most recent count the AKC registry recognised 202 different breeds, while the Kennel Club total is slightly higher, at 218. The two clubs also assign breeds to one of a small number of descriptive classes: gun-dog, hound, pastoral, terrier, toy, utility and working in the UK, and very similar classes in the USA. There are kennel clubs in almost every country and most are affiliated to the Fédération Cynologique Internationale, an umbrella body formed in 1911 with headquarters in Belgium. Their stated remit is equally explicit: '... to encourage and promote breeding and use of purebred dogs whose functional health and physical features meet the standard set for each respective breed and which are capable of working and accomplishing functions in accordance with the specific characteristics of their breed.'

Kennel Club breed classes

Class	Typical breeds
Gun-dog	Retrievers, Spaniels, Pointers, Setters
Hound	Scent: Beagle, Bloodhound
	Sight: Whippet, Greyhound
Pastoral	Collie, Sheepdog, Samoyed
Terrier	Highland, Jack Russell
Toy	Chihuahua, Pomeranian
Utility	Bulldog, Dalmatian, Akita, Poodle
Working	Boxer, Great Dane, St Bernard

New breeds are added to the list from time to time and there are strict criteria for their acceptance. Every breed has an agreed written standard to which pedigree dogs must conform. These are mainly concerned with their appearance, with temperament coming a distant second. They are policed by dog show judges who use the breed standard as a yardstick for scoring competitors.

This is a typical Kennel Club breed standard, in this case for the Beagle:

> *General Appearance*
> A sturdy, compactly built hound, conveying the impression of quality without coarseness.

Characteristics

A merry hound whose essential function is to hunt, primarily hare, by following a scent. Bold, with great activity, stamina and determination. Alert, intelligent and of even temperament.

Temperament

Amiable and alert, showing no aggression or timidity.

Head and Skull

Fair length, powerful without being coarse, finer in the bitch, free from frown and wrinkle. Skull slightly domed, moderately wide, with slight peak. Stop well defined and dividing length, between occiput and tip of nose, as equally as possible. Muzzle not snipy, lips reasonably well flewed. Nose broad, preferably black, but less pigmentation permissible in lighter coloured hounds. Nostrils wide.

Eyes

Dark brown or hazel, fairly large, not deep set or prominent, set well apart with mild, appealing expression.

Ears

Long, with rounded tip, reaching nearly to end of nose when drawn out. Set on low, fine in texture and hanging gracefully close to cheeks.

Mouth

The jaws should be strong, with a perfect, regular and complete scissor bite, i.e. upper teeth closely overlapping lower teeth and set square to the jaws.

Neck

Sufficiently long to enable hound to come down easily to scent, slightly arched and showing little dewlap.

Forequarters

Shoulders well laid back, not loaded. Forelegs straight and upright well under the hound, good substance, and round in bone, not tapering off to feet. Pasterns short. Elbows firm, turning neither in nor out. Height to elbow about half height at withers.

Body

Topline straight and level. Chest let down to below elbow. Ribs well sprung and extending well back. Short in the couplings but well balanced. Loins powerful and supple, without excessive tuck-up.

Hindquarters

Muscular thighs. Stifles well bent. Hocks firm, well let down and parallel to each other.

Feet

Tight and firm. Well knuckled up and strongly padded. Not hare-footed. Nails short.

Tail

Sturdy, moderately long. Set on high, carried gaily but not curled over back or inclined forward from root. Well covered with hair, especially on underside.

Gait/Movement

Back level, firm with no indication of roll. Stride free, long-reaching in front and straight without high action; hindlegs showing drive. Should not move close behind nor paddle nor plait in front.

Coat
Short, dense and weatherproof.

Colour
Tricolour (black, tan and white); blue, white and tan; badger pied; hare pied; lemon pied; lemon and white; red and white; tan and white; black and white; all white. With the exception of all white, all the above-mentioned colours can be found as mottle. No other colours are permissible. Tip of stern white.

Size
Desirable minimum height at withers: 33 cms (13 ins). Desirable maximum height at withers: 40 cms (16 ins).

In summary, this is a pretty stringent set of criteria to which breeders, if they show their dogs, try hard to comply.

From a genetic point of view, modern pedigree breeds conforming to Kennel Club rules are the canine equivalent of the desert island we introduced back in Chapter 5 from which no one leaves and at which no one arrives. Once on the island, they are completely cut off from all other dogs. They can only breed with each other and this is the perfect scenario for inbreeding. Strange things can happen with inbreeding in any species, and to understand them we need to go back to the basic biology. I will borrow some of my examples from human genetics, as we understand these very well indeed.

The problems of inbreeding all stem from a fundamental aspect of mammalian biology. They are a consequence of having not one but two pairs of each chromosome. This is the case for all mammals including, of course, dogs and humans. The genes that we and all mammalian species need in order to live and reproduce are located on these chromosomes. One of each pair comes from the mother and one comes from the father.

Thus we have a paternal copy and a maternal copy of each gene. Different species have different numbers of chromosome pairs, twenty-three in humans and thirty-nine in dogs. The number is relatively unimportant. It is the genes on the chromosomes which really count. Ideally we, and dogs, need all the genes from both parents to function properly, but we can usually get by with one, as long as it is in good condition. Under these circumstances the genes on the other chromosome act as back-up.

A harmful mutation in a gene on one of the chromosome pairs is usually offset by the normal, fully functional gene on the other member of the pair. However, if the 'normal' chromosome cannot compensate for the faulty mutant, then the individual will suffer from a genetic disease which will, almost always, reduce the prospects for the individual to have offspring. In humans this usually means that affected individuals have fewer children, while in dogs the mutants will not be chosen for breeding and may be euthanised, unless the mutation happens to be a Darwinian 'sport' with exotic appeal.

When the 'normal' chromosome is able to compensate fully for its damaged partner, the individual, human or dog, will usually show no outward sign of the problem that lurks within. Such an individual is known as a 'carrier'. It is a problem that only surfaces in later generations, if two carriers meet and breed. The simple rules of genetics dictate that when this occurs, one half of pregnancies will be carriers like their parents, one quarter will have two 'normal' chromosomes but in one quarter *both* chromosomes will carry the mutant gene. Without a normal gene to compensate for the loss, the individual will show symptoms, whatever they happen to be. The disease is then said to be *recessive*.

Within human communities, genetic diseases which only show when both parents are carriers have been known for a long time and are the main reason for restrictions on the marriage of biologically close relatives that are widely taught by many religions. By definition, close blood relatives share a common ancestor in the recent past. Full siblings share both parents as ancestors, first cousins have the same grandparent and so on. Taking the last example, if the shared grandparent is a carrier of a genetic disease, he or she would be completely normal and almost certainly be unaware of being a carrier. However, the mutant gene may be passed on to his or her descendants, again following the simple rules of inheritance. The parents of the two cousins, who are siblings, each have a 50 per cent chance of inheriting the mutant gene from whichever of their parents was the carrier. That gives them a 25 per cent chance that *both* are carriers. If they have children, every pregnancy will have a one quarter chance of inheriting the mutant gene from both parents. With a double dose of the damaged gene, and lacking the back-up of the normal counterpart, the child will exhibit all the symptoms, sometimes with devastating consequences.

The reason that this type of inherited disease is rare in outbred communities is that, though all humans are symptomless carriers for around fifty serious genetic diseases, the chances of their descendants marrying one another is small. However, some genetic disorders are far from rare. The commonest human recessive disease in Europe is cystic fibrosis, which is usually caused by a single mutation in a protein within the cell membrane that regulates the passage of chloride ions in and out of the cell. The unfortunate cystic fibrosis patients produce a much stickier mucus than normal and this builds up in the lungs. Although tedious daily physiotherapy can help clear the mucus, there is a greatly increased risk of lung infections, and most cystic fibrosis patients die of pneumonia before they reach forty.

The proportion of cystic fibrosis carriers in Europe is surprisingly high, at one in twenty. The reasons for this are extremely interesting, and we will deal with them a little later. But, again following the simple rules of genetics, if the carrier rate is one in twenty, the chances in an outbred community of two carriers becoming parents is 1/20 x 1/20, that is 1/400. Under these circumstances each pregnancy has a one in four chance of inheriting two copies of the mutant gene, giving an overall prevalence of cystic fibrosis of 1/1600.

If, however, the two parents are descended from a recent shared carrier ancestor, the incidence of the disease increases dramatically. This is especially notable in some closed religious communities, like the Old Order Amish in Lancaster County, Pennsylvania. The ancestors of present-day Amish, and the closely associated Mennonites, emigrated from Germany in the early eighteenth century. They are certainly prolific and, by 2012, their numbers had grown to almost a quarter of a million. The Amish have an unusual way of life, eschewing modern amenities like cars, telephones and electricity. They drive around in horse-drawn carriages, and if you have ever been taken for a ride around Central Park in New York, the carriage was probably made by an Amish craftsman.

I mention the Amish because in many ways, from a genetic point of view, they resemble the closed breeding communities of the pedigree dog. The Amish only reproduce within their own communities, just as pedigree dogs must do if breeders want their litters to be registered. One might imagine that the incidence of genetic disease in the inbred Amish would be much higher than in the general outbred population, but this is not so. They do, however, have higher rates of particular genetic conditions, some of them serious. One of these is SCID or severe combined immunodeficiency disease, also known colloquially as 'bubble boy syndrome'. SCID patients have such a weakened immune system that they must be completely isolated from all sources of infection until they are old enough to receive a bone marrow transplant. SCID among the Amish is caused by a mutation on chromosome 15 called IL7R, which is a receptor for interleukin, a protein concerned with inter-cellular communication. Mutations in other genes also cause SCID but not in the Amish, so it is only the specific IL7R mutation that has been inherited from the shared carrier ancestor of affected Amish children. It was almost certainly present in one of the two hundred original founders of the Old Order

Amish who emigrated from Germany. It remains hidden within the population until exposed in a double-dose SCID patient.

Similarly, pedigree dogs, as we have seen, come from a genetically limited pool of founder ancestors. If one of their number is a carrier of a mutant gene it is in the gene pool and will lie dormant until it shows itself when two mutant genes are combined within a single individual. However, if conversely there are no carriers among the founding dogs, the whole pedigree will be free of the disease.

Theoretically, now that we can identify the precise mutation, the Amish could all be tested to see who was a SCID carrier and avoid the situation altogether by making sure two carriers did not marry or at least did not have children. However, their religious views forbid them from participating in preventative genetic testing, instead accepting the diseases as 'God's will'. Other human communities have found themselves in a similar position but have shown a determination to do something about it.

For example, another fatal inherited neurological disorder, Tay-Sachs disease, affects the Ashkenazi Jewish community of New York and elsewhere. They are not quite such a completely closed breeding community as the Amish but they do show a high degree of inbreeding as a result of their tightly knit East European origins. They used to have a high rate of Tay-Sachs disease for the same underlying reason as the Amish, again because inbreeding often brought together two carrier parents. The Tay-Sachs mutation is located on chromosome 15 in a gene called HEXA. Identifying the genetic mutation means that carriers can be identified by a simple DNA test. Unlike the Amish, the Ashkenazi community have enthusiastically embraced DNA-based carrier screening. Carriers are discouraged from marrying other carriers and, as a result, the incidence of the disease has dropped to zero. There are now more Tay-Sachs babies born to non-Jews, who are at very low risk so aren't screened for carriers.

There are strenuous efforts among pedigree dog breeders and organisations like the Kennel Club to emulate the success of the Ashkenazi

Jews in eliminating Tay-Sachs disease by using DNA tests to screen pedigree dogs to reveal carriers of diseases common to their breed and discourage them from breeding. This will quickly reduce the frequency of the disorder. In time, the incidence of carriers within the pedigree breed will go down and, with sufficient effort, could be eliminated altogether. From this point onwards the breed would be free from the disease.

In the UK the leading testing lab, the Kennel Club-sponsored Animal Health Trust (AHT), has developed tests for twenty-two different inherited disorders present in sixty-five different dog breeds. Other centres have also developed DNA tests for carriers and breed identification, of which more later. The same mutation, and the same disease, can be present in different breeds because some of the dogs that were used to create the breeds were already carriers. The Trust laboratories are near Newmarket in Suffolk and boast an impressive track record. They have DNA tested over 85,000 dogs from around 50 different countries for the mutant genes responsible for the 22 recessive disorders and identified nearly 10,000 carriers.

Each of these mutations, in common with all genetic mutations, has arisen spontaneously. Where the mutation and the disease are found in only one breed, the likelihood is that they arose after the breed was closed. However, as I mentioned above, when a specific mutation and its associated disease occur in more than one breed, the chances are that the mutation arose before the breed was closed to outsiders.

I arranged a visit to the AHT labs and found out more about their activities and a lot more besides, which we will come to in Chapter 20.

The Dog Genome

At a ceremony on 26 June 2000 a group of distinguished scientists and public figures gathered in the East Room of the White House to hear President Clinton announce the completion of the human genome DNA sequence. It was a day of celebration and optimism after a long and sometimes unseemly race between the public and private sectors to complete the sequence. The publicly funded initiative was led by Francis Collins, director of the Human Genome Project, while the private effort was represented by Craig Venter, president of the privately funded biotech company Celera Genomics. The celebration was well deserved. Scientists from many countries had contributed to revealing, in their correct order, the 3 billion DNA units that make up the human genome.* The sense of optimism was tangible but, unfortunately, less well founded.

President Clinton, in his address, predicted, 'It will revolutionise the diagnosis, prevention and treatment of most if not all human diseases.' An armada of projects committed to turning this dream into a reality was launched around the world. The possibility of failure was never contemplated when President Clinton predicted that, in the near future, cancer would be known not as an incurable disease but 'only as a constellation

* Technically this was the first draft of the human genome sequence, substantially complete but excluding some stretches of unimportant 'junk DNA'.

of stars' in the night sky. Tragically, this has not turned out to be the case and cancer is just as much a scourge today as it ever was.

Although most diseases have a genetic basis, unfortunately very few of them are caused, like cystic fibrosis, SCID or Tay-Sachs, by mutation of a single gene. The diseases that inflict the most damage on the most people have generally turned out to involve a multitude of genes, each of small effect. Without clear gene targets to attack, the scientific armada was soon drifting aimlessly. The blame for the underestimation of the wretched complexity of the genome was laid squarely at the feet of our own stubborn refusal to behave as laboratory animals. As Alfred Sturtevant, one of the early pioneers of genetic mapping, remarked, progress in human genetics was severely limited because 'unfortunately, breeding experiments with humans are generally frowned on'. Despite this worrying statement, Sturtevant was actually a vociferous opponent of the eugenics movement sweeping Europe and America in the early twentieth century. He chose the fruit fly *Drosophila* as his model animal. Others have used mice or zebra fish, but although these species have admirable qualities as laboratory animals and can be genetically manipulated to harbour mutations, their patently obvious dissimilarity to ourselves had frustrated researchers who had hoped to use these species to identify human disease genes. At that point, scientists realised that they might be able to speed up the process of human disease gene identification by using a much more familiar animal. Dogs suddenly became the newest destination for the armada.

All armadas need charts to guide them, and a few years after the White House announcement the sequence of the dog genome became the focus for a powerful team of genetic cartographers. Their headquarters was the newly opened Broad Institute on the banks of the Charles River in Boston where it was closely allied with the nearby MIT and Harvard. Hugely well endowed, the Broad is awash with the very latest equipment and top-notch scientists to match. At the helm of the institute is one of the doyens of human genome research, Eric Lander, with several scientific scoops already under his belt. The Broad was keen for a substantial success to merit its extravagant funding, and the dog genome was to be it.

Lander and his team had the job done quickly and the dog genome was published in full in *Nature* on 5 December 2005.[1] It is crammed with detail, much of it beyond the scope of this book, and introduced several new ways to analyse the vast amount of data generated by the relentless grinding of the DNA sequencing machines. The dog was the fourth complete genome to be published, after the human, mouse and rat, the last two being obvious targets as popular laboratory animals. In overall size, the dog genome is smaller than that of the human by about 500 million bases (2.8 versus 3.3 gigabases, abbreviated as Gb) and, perhaps surprisingly, smaller than that of the mouse. But genome size is not a good guide to complexity and, as I enjoyed telling my students, the size of the human genome lies between those of the lupin and the newt.

There are also fewer dog genes, the parts of the genome that encode proteins. By searching out the tell-tale DNA sequences marking the beginning and end of genes, the Broad team found 19,300 protein-coding genes in the dog, compared to 22,000 in humans.

Another type of analysis revealed the parts of the genomes of both dog and human that had been evolving quickly. Earlier findings had shown that, compared to the mouse, the most rapid genome evolution in humans was to be found in genes concerned with brain development. This fitted perfectly with our vain self-image of intellectual superiority within the animal kingdom. The dog genome, however, showed an equally rapid rate of evolution in dog brain genes. Interestingly, the other genes that appear to have accelerated evolution in humans were associated with sperm production and mitochondria.

The latter are not the genes carried on mitochondrial DNA itself, but on mitochondrial genes that have been 'kidnapped' by our nuclear DNA over the course of evolution. I found the explanations put forward for this observation most intriguing. They suggest the powerful influence that sexual selection has had on our own evolution. Competition between sperm for the prize of success in fertilisation is well known across many species. It is especially intense in primates and may well have encouraged the rapid evolution of genes conferring the production of greater quantities of sperm or speedier motility to beat the competition to the egg. Less clear-cut is the rapid evolution of the kidnapped mitochondrial genes themselves, though a case for sexual selection can be made here as well, bearing in mind that sperm are powered by mitochondria as they paddle furiously towards the finishing line, marked by the unfertilised egg.

Genomes are not sequenced in their entirety all at once, but in chunks containing roughly 50,000 base pairs. The sequenced segments must then be assembled correctly to recreate the DNA sequence as it is in the genome. Chromosomes are simply long, very long, linear strings of DNA. On each chromosome of a pair the order and position of genes along it is the same, but the precise DNA sequence is slightly different in ways we will look at soon. It is no small task to place the sequence of chromosome segments in the correct order. This is achieved by sequencing multiple DNA segments that overlap one another. Powerful, very powerful, computers then match up the overlaps and produce the complete sequence. Such is the complexity of this process that it comes as no surprise that there are so many software engineers among the forty-four authors on the 2005 paper that announced the complete sequence. Forty-four authors, but only one dog, contributed to the project – a female Boxer called Tasha.

The Broad paper trumped an earlier partial dog genome sequence produced by Craig Venter and his team at the privately funded Institute of Genome Research (TIGR) in Maryland. As already mentioned, Craig Venter led one of the teams involved in the original discovery of the human genome sequence. In the long-standing tradition of scientists using themselves as guinea pigs, it was Venter's own DNA that was the first to be sequenced. And it was Venter's own poodle, Shadow, that became the very first dog to have its genome laid bare by science.

In addition to sequencing the dog genome, and thoroughly checking for accuracy, the Lander team at the Broad Institute also searched Tasha's genome for bases that differed between the chromosomes in a pair. These could then be used, in ways we shall look at shortly, to trace genetic disease traits, at first in dogs and later perhaps in humans too. The first search examined the detailed sequence of Tasha's two chromosomes for differences involving just a single base pair. This kind of difference, caused originally by a mutation during DNA copying, is perpetuated in subsequent generations and can spread throughout many of the descendant dogs.

Lander's team identified all the positions in the dog genome where the two chromosomes differed at a single base. For example, there might be a T base on one of the pair and a C base at the equivalent position on the other. These variations in the sequence are known as single nucleotide polymorphisms (SNPs, pronounced 'Snips'). Just by comparing the sequence of Tasha's two chromosomes, the team from the Broad Institute found an astonishing 770,000 bases at which the two chromosomes differed. Once the genome sequence of Venter's poodle Shadow was included to enhance the search for SNPs, the total increased the count to nearly 1.5 million. By a bit of rough and ready partial sequencing of nine different dog breeds the number of SNPs went up to 2.5 million. That is an astonishing total but still only averages out at one SNP per 900 DNA bases, or just over 0.1 per cent, across the dog genome.

There is a lot that can be done with 2.5 million SNPs, especially when you know exactly where they are within the genome. Firstly, they can be used to manufacture 'chips' that can rapidly test any dog's DNA for all 2.5 million of them. In practice this is more than enough, so the chips that were manufactured were restricted to a more manageable, and economic, 100,000 SNPs.

To explain the value of these markers I need to introduce one vital aspect of chromosome behaviour. We've covered the fact that in mammals all individuals have matched pairs of chromosomes, one coming from the mother, the other from the father. Each carries the same genes in the same places, and for most of their life in the cell nucleus, the chromosomes keep themselves to themselves. The genes issue instructions to the cell that are followed to the letter and synthesise the whole range of proteins that we and all other animals need to grow and sustain life.

Blood genes tell red blood cells how to make haemoglobin, bone genes tell bone cells how to make collagen, hair genes tell hair cells to make keratin and so forth. All the while the sets of genes on the two chromosomes act independently of each other, quietly getting on with the job of life. Meanwhile, a few cells, inside the testis and the ovaries, are being made ready for the next generation. These are the germ cells, to distinguish them from the workaday somatic cells, and their sole function is to pass on their DNA to the next generation. But not all of it. As we have seen, somatic cells contain pairs of chromosomes, thirty-nine pairs in the dog and twenty-three pairs in humans. Germ cells, however, that's to say sperm and eggs, only have one of each pair. When a sperm fertilises an egg the two sets of chromosomes are combined and the normal number is restored. But something else happens too. Within the germ cells that go on to become sperm and eggs the pairs of chromosomes begin to dance with each other, moving closer and closer until they are touching. At these fleeting contacts something truly amazing happens. The chromosomes, those long strands of DNA, actually break and re-form with their dance partner. The embrace is short-lived and is over in a matter of seconds. The entwined chromosomes break the clinch and move apart. But during that brief embrace something truly amazing occurs, as we will see a little later.

17
The Genetics of Pedigree Breeds

Once the dog genome and the vast range of SNP markers were made available in 2005, scientists were quick to capitalise on the bonanza by re-drawing the evolutionary tree of dog breeds constructed with mitochondrial DNA over twenty years previously. In 2010 *Nature* published the results of a comprehensive study of 912 dogs from 64 breeds.[1] The authors were a group of thirty-six scientists led by veteran dog biologist Robert Wayne and Bridgett M. vonHoldt. Reassuringly, all branches of the resulting tree led back to the wolf, ruling out any major contribution to the dog genome from other species that, theoretically at least, might have entered the gene pool through the male line. See a visual representation of this tree, or phylogram, on the next page.

The relationships defined in this tree are based on the overall similarity between the sequences of the autosomes, the thirty-eight pairs of dog chromosomes not involved in sex determination. Breeds that cluster together share more identical DNA sequences than breeds that are far apart. Unlike the mitochondrial tree (Wayne/Vilà, page 17), the autosomal tree is drawn in a circular pattern. Breeds are displayed around the perimeter and the deduced links between them are towards the centre. The branching order is a rough approximation of the time that has elapsed since the breeds were established. From there, branches radiate out from the wolf, first to the so-called 'ancient' breeds, the Basenji,

Australian Dingo and Chow Chow, then the Asian and Arctic spitz breeds. The next branch leads to the Afghan and Saluki, and at the end of the next branch are the Samoyed and American Eskimo dog, grouped closely together as one might expect with the Arctic breeds. After these, the next major limb of the tree carries all other breeds.

This is a good point at which to remind ourselves that, unlike thoroughly researched human family trees for example, where all the links are known, these trees represent the *most likely* rather than *absolute* scenarios. We are, after all, drawing them as best we can to fit the genetic data. The absolute accuracy of the tree can never be guaranteed. Neither is it strictly

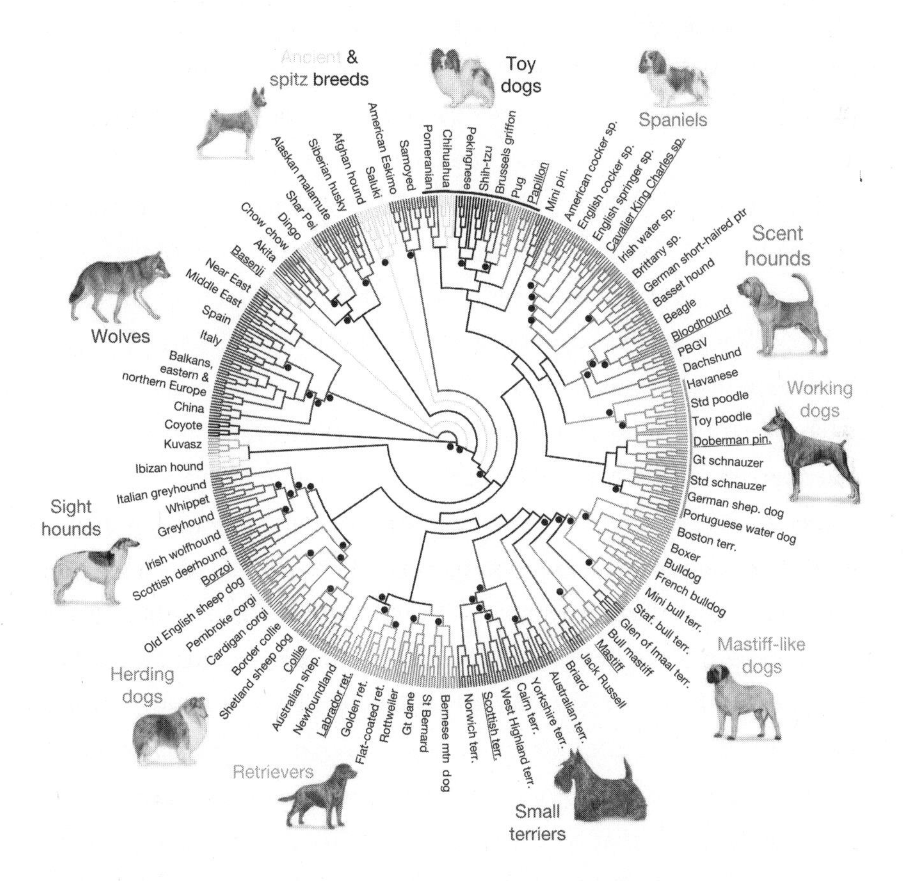

This example of a phylogram, based on Wayne and vonHoldt's 2010 study, illustrates (autosomal) genetic similarity in a range of dog breeds.

speaking an evolutionary tree, though there are evolutionary inferences to be drawn from it. Although it might look like a real family tree that starts at the centre and grows outward, it is really a diagrammatic representation of the genetic similarities between different breeds of dog. It is reasonable to assume that breeds on two close branches might have originated from a common founder, but it may not be quite so simple. So, for example, although the Samoyed and American Eskimo dog are placed at the tip of a long branch it doesn't mean that all the dogs of both breeds are descended from a single 'common ancestor'. As we shall see in a moment, there has been a great deal of mixing between breeds that we know about but which doesn't show if we treat the diagram like a genealogy. This kind of diagram is formally known as a phylogram rather than a tree in an attempt, not always successful, to avoid the confusion with true genealogies.

There are ways, which I won't trouble you with, for getting the tree that 'best fits' the data, and that is what is reproduced on page 117. The same algorithms also produce alternative trees that don't fit quite so well, perhaps differing in the detail of some of the branches. These might be closer to reality, but not by much.

The most revealing aspect of the autosomal tree is the way that most breeds group with others of the same type as defined by the UK and US kennel clubs. Scent-hounds group with scent-hounds, mastiffs group with mastiffs, herding dogs with other herding breeds and so on. This is quite different, if you recall, from the tree derived from mitochondrial DNA, where individual dogs of the same breed are often descended from different ancestral females. This isn't quite as surprising as it sounds at first. Although mitochondrial DNA contains the genes that enable cells to convert food to energy using oxygen, these are only a small fraction of the thousands of genes that do everything else. The fact is that the individual dogs of the same breed have different mitochondrial ancestors, and also that dogs of different pedigree breeds can share the same mDNA ancestry.

It is a reflection of the fact that most of the features that distinguish breeds, like appearance, temperament and so forth, are controlled by many other genes. It is the similarity in all these other genes that explains the remarkable clustering not only of breeds but also of breed type in the genome phylogram. Thinking of the phylogram as a summary of similarities in all genes, it becomes less of a surprise to find similar breeds and breed types clustering together.

During the work leading up to this definitive summary, the question arose as to whether genetic analysis could be used to predict the breed of a dog from its DNA alone. This was first attempted back in 1999 in a case before a German county court. The essence of the case was that a car collided with a dog and suffered substantial damage to its front end. Completely disregarding its legal obligations, according to the police report, 'After the collision, the dog left the scene of the accident without proving his identity ...'

The driver and owner of the car filed a law suit including a claim for damages against a local shopkeeper who he suspected was the owner of the dog. One of his two German Shepherds had apparently been treated for minor injuries around the time of the accident. The judge ordered a test on the injured dog to compare its mDNA to three hair fragments recovered from the damaged car. The sequences from the hairs left on the car and the defendant's dog were both definitely canine but they differed from each other in their detailed sequences. After this result was brought to court, the case was dismissed and the suspect walked, or at least limped, free. Although the mDNA analysis cleared the dog, it was only later that it was appreciated that mDNA could not have been used to differentiate the breeds. For that, it was essential to have nuclear DNA sequences.

Heidi Parker in Robert Wayne's lab was the first to explore in detail the clustering of breeds using nuclear genes, and in 2004 she published her results in *Science*.[2] This was the year before the dog genome sequence appeared, with its galaxy of SNP markers, and she had to rely on a different but still effective genetic system based on what are known as microsatellites. These are small segments of DNA that occur as repeated blocks a few bases long. They are very useful as genetic markers because the number of blocks in a run can vary and is comparatively easy to measure. Genetic fingerprinting, invented by the British geneticist Alec Jeffreys in the late 1980s, is based on microsatellites.

Parker and her colleagues used a panel of 96 microsatellites in 414 dogs from 85 different breeds, and tried to assign the breed from the genetic data alone. It was a remarkably successful exercise, assigning more than 99 per cent of 'test' dogs to the correct breed. Only four dogs were classified incorrectly: a Beagle was identified as a Perro de Presa Canario; a Chihuahua, as a Cairn Terrier; and two German shorthaired pointers, one as a Kuvasz, the other as a Standard Poodle. The correspondence was quite remarkable, with 410 of 414 assignments being spot-on, but it was nonetheless surprising that the 'errors' assigned dogs to apparently unrelated breeds rather than something rather similar.

By 2004 genetic genealogy and ancestry testing for humans was well under way and it was no surprise that companies saw an opportunity to put the technique to good use in dogs. Not only might it appeal to the owners of purebred dogs, but it could also offer a way of determining the mixed ancestry of mutts. As with any commercial operation there is a balance to be struck between accuracy and price. There are no rules governing the fidelity of such tests, as there are for human health diagnostics, which must gain approval by the US Food and Drug Administration (FDA), for example. Testing companies compete on reputation and price just as they do in clothing, cosmetics and a thousand other consumer products. The science behind the tests is based on the type of analysis which Heidi Parker and her colleagues pioneered with the pedigree breeds. However, it does not follow automatically that the remarkable accuracy in breed assignment achieved by Parker with pedigree animals will translate seamlessly into a comparable accuracy when it comes to mixed breeds. Let me explain.

Parker's original 2004 assignments were based on genetic similarities of a number of microsatellite genetic markers, while the more advanced 2010 treatment used SNPs. None of the markers in either system is diagnostic of the breed on their own. It is only when they are combined that the 'most likely' breed assignment can be made for individual dogs.

I've been testing DNA, in humans, for more than twenty years and am struck by the reputation for invincibility that DNA has acquired and has maintained over all these years. I think it began after the remarkable precision of genetic fingerprinting and its rapid and well-publicised application to some very grisly cases of rape and murder. The stunning accuracy of individual identification claimed for the technique, often to the level of one in several billions, was thoroughly and publicly tested in the courts. DNA secured convictions of the guilty and overturned wrongful incarceration of the innocent. Faced with the evidence, rapists changed their pleas to guilty and victims were spared a courtroom cross-examination. This is genetics at its most impressive.

The years have only enhanced DNA's reputation for invincibility, and it has now become part of everyday vocabulary. Not in the original sense of the word (it isn't a word, incidentally, but an acronym – best forgotten – for Deoxyribo Nucleic Acid), but as a metaphor for a mysterious essence. This is great news for geneticists like me who find themselves wrapped in the cloak of invincibility and assuming the Delphic power of the Oracle. But, in life as in myth, it only goes so far. Although a DNA sequence is an ultimate truth of a sort, it needs an all too human oracle to interpret it. And like the classical example, the exaggeration of its power is a perennial temptation.

In my own work on human ancestry, expectations are often unrealistically high. One woman was genuinely astonished that I was unable to tell her, from a DNA test on her mother, why her second cousin from

Yorkshire had freckles. Another complained when I told him he probably had Celtic ancestors. 'I know that already,' he said. 'How come?' I asked. 'Because I have dark hair and blue eyes.'

Returning to dogs, it's my impression that many owners who have their mixed breed dog's DNA tested expect a similar ultimate truth. Companies are understandably wary of disabusing their customers of the power of their product, and that makes unrealistic claims all the more tempting, especially in an intensely competitive market, and I have found it quite difficult to understand the claims of competing companies. It's fun to have your dog's ancestry tested, but bear in mind that there are limits to the accuracy of these tests, for the reasons we have covered. And the more mixed the ancestry, the more inaccurate the assignments. In my very skimpy reading of available tests I can scarcely believe that a proper breed assignment test can be sold for as little as £10.

I thought it was time I tried one of these tests and asked Ulla to look out for a mixed breed dog whose owners would be prepared to volunteer their dog. This was easier said than done. As we will see later in the book, Ulla interviewed several owners who regularly walked their dogs in London's Hyde Park. Because it's an affluent area, all the dogs that she met in the park were pedigree specimens and I was beginning to think we would never find a mutt to test. At the very last minute, only three weeks before the book manuscript had to be with the publishers, Ulla came across Archie in a local pub, along with Chris, the landlord, his wife Helen and their 10-year-old son George, who took the saliva sample. Archie was not a complicated mongrel; he was what is called a 'designer' dog. He was a Labradoodle, a cross between a Labrador and a Standard Poodle. They specifically wanted a Labradoodle to guard against allergies.

The DNA sample was sent to what looked like a reliable testing lab to see what they were able to deduce about Archie's genetic composition. The results were turned around with admirable speed, so there was time to ask the family what they made of the genetic results. Their first reaction was one of surprise. The DNA test revealed that Archie was a mixture of 62.5 per cent Standard Poodle, 12.5 per cent Labrador and 25 per cent Golden Retriever. Archie is jet black, so to find out that he is one quarter Golden Retriever came as a bit of a shock. I am confident that the results were technically accurate insofar as they reflected the best fit of Archie's DNA to the company's extensive database on pedigree breeds, and I have no intention of challenging the report's conclusion. The more interesting question for me is this. Is Archie a mixture of five-eighths Standard Poodle, one quarter Golden Retriever and one eighth Labrador, as the DNA results tell us, or is he a Labradoodle, which is what the breeder told Chris and Helen when they bought him?

'I suppose I'll have to go with the science,' replied Chris, a little reluctantly. Ulla then enquired whether, had he known the results of the genetic test, he would still have bought the dog. 'Had I known that, I wouldn't even have driven all that way to see him.' But once they saw Archie they were hooked.

Labradors and Golden Retrievers are genetically very close. We can see that from the phylogram on page 117 where the two breeds occupy adjacent positions. Also, remembering how the Golden Retriever breed was developed by Lord Tweedmouth, one of the ancestors of Queenie, the original Golden Retriever, was Sambo the Labrador. Even if the DNA conclusions were not quite right, they are very close to what was expected. Assuming that the breeder is being completely honest about Archie's pedigree, and there is no reason to doubt that, is he a Labradoodle or is he not?

Pedigree breeds are defined by the breed standard and not by genetics, at least for the moment. Although Archie is a crossbred dog, his two parents, one assumes, were a pedigree Labrador and a pedigree Poodle, each with its own breed standard. From a purely genetic point of view, that cannot be true. The simplest genetic explanation for the one-quarter Golden Retriever in Archie's make-up is that the equivalent of one of his grandparents was a Golden Retriever. My own feeling is that the explanation is neither of these two apparently conflicting options. Instead, our judgement is dulled by the drowsy syrup of Horus, the Egyptian god of numbers. Once complex issues are reduced to numbers, we seem to abandon our sense of judgement. Genetically, there is no such thing as a 'pure' dog breed any more than in human genetics there is such a thing as a 'pure' race or a 'pure' ethnic group. In humans, this fallacy has led to dangerous misunderstandings and harmful discrimination. I doubt the fallacy of 'purity' will infect the dog world, but it is best to be on guard against its malign influence. It certainly does not bother Archie's owners. Once he had recovered from the surprise about Archie's part Golden Retriever heritage 'revealed' by the DNA test, Chris reached down to stroke his neck. He was still the same dog. He was still Archie. Helen and George smiled in agreement.

The Dance of Life

We have seen how by comparing overall similarities in DNA sequence we can reveal genetic relationships between different breeds that make sense. This can be taken to an even higher level by taking advantage of how chromosomes are organised. To see how, we return to the dance of the chromosomes we first encountered at the end of Chapter 16.

As the germ cells, the sperm and eggs, are formed, something very important occurs. After the music stops, the chromosomes move apart to continue their lives in isolation, waiting for the next time they can take to the floor. For the great majority of germ cells that time never comes. Only the chromosomes in the cells that go on to be fertilised will ever dance again, and for that they must wait until the next generation. The rest are discarded.

Under the microscope the chromosomes look the same after the dance as before. But, despite appearances, they certainly are not. During the brief encounter the chromosomes have broken apart, exchanged segments of DNA with their dancing partner and re-formed in new combinations. Each chromosome pair breaks and re-forms at a slightly different place so that, after the embrace, there are an almost infinite number of chromosome combinations in the germ cells, each one a different mosaic of paternal and maternally derived segments.

The whole point of this shuffling, and indeed the whole point of sexual reproduction itself, is to create genetic variability in the offspring. It's this shuffling that makes sure that a child, or a puppy, never has exactly the same genetic make-up as his or her parents. There is a very good reason for going to all this trouble. We are, and have been throughout evolution, the target of pathogens. Bacteria, viruses, fungi and parasites of all sorts are constantly looking for an opportunity to invade our bodies. Were it not for our defences, we would soon be overwhelmed. Our immune system is the main defensive bulwark against pathogen invasion and it works silently and tirelessly to eliminate these threats from our system. Only when it is broken, by AIDs or immuno-suppressive drugs, do we appreciate how well our immune system has defended us, and how vulnerable we are to infection without it.

Pathogens themselves counter this solid defence by changing their genetic make-up. This is the reason for antibiotic resistance, for example, where a single bacterium mutates and is no longer killed by the drug. It multiplies and prospers among the dying legions of its contemporaries, and the infection takes hold again but this time it cannot be stopped by the antibiotic. Our defence against the pathogens' counter-attack is to change our own genetic make-up at every generation, and we do this by gene shuffling.

Imagine splitting a pack of cards with just two suits of different colours, say hearts and spades, into separate piles with hearts in one, spades in the other. The order of the cards is the same in both, say ace to king in the usual way. These two decks represent a single pair of chromosomes before recombination. The order of genes is the same on both, but there are slight differences in their sequences. Now we cut both decks at the same place and swap them over. The numerical order of the recombined decks remains the same: Ace, two, three etc., to King, but at the cut, the colour changes from black to red in one pack and from red to black in the other. This represents the shuffling of one pair of chromosomes in one generation. There are already a lot of different possible combinations in the final decks and this number increases as we add in the other chromosomes. Even with this crude analogy it is easy to appreciate that the total number of possible combinations is vast. The shuffled chromosomes separate into different germ cells. Thinking of sperm, most will perish, but one will survive to fertilise an egg where it meets up with the shuffled deck from the mother. Repeated over the whole set of chromosomes, the possible combinations of genes are mind-bogglingly huge. Almost certainly that exact combination will not have been seen before in the whole history of a species. Importantly, and this is the point, pathogens will not have seen the same genetic combination before and will not have time to evolve a way of overcoming it.

Some plants, and a few animals, have done away with sexual reproduction altogether. The strategy can be very successful because they no longer need two sexes. Males are redundant. The glory is, however, short-lived. Sooner or later a pathogen will unlock the key to the defence system and, once accomplished in a single individual, will sweep through the entire population with devastating effect. As an example, South American bananas are propagated by cuttings and not sex, so that all banana plants are genetically identical clones. A fungus, *Fusarium oxysporum*, managed to unlock the defences in one plant, and is now sweeping through the entire crop.

I have taken rather a long time to explain the evolutionary reasons behind gene shuffling, but there is a very practical effect. Gene shuffling, or recombination to give its proper title, means that we can map and locate genes. In dogs, these might determine size, or coat colour, or tail shape, or perhaps a genetic disease like hip dysplasia. This is how it works.

Taking the risk of overworking the playing card metaphor, let's suppose we want to find the gene responsible for one of these traits, as they are called. As a relevant dog example, we will take the form of hip weakness that affects several related pedigree breeds, like Labradors. We may think it's caused by a faulty gene, but we have no idea which one. Sequencing the entire dog genome will indeed give us the sequence of DNA of the dysplasia gene, along with all 19,000 others. From this alone, there is no way of telling which of these genes is associated with the hip problem. You can make a stab at it by examining the sequence of genes that *might* be involved, like cartilage collagen genes, say, but it's just a guess and may well be wrong. It usually is.

This is where recombination comes to our aid. Let's say that, without looking, we insert a joker, representing the dysplasia gene, somewhere in the deck. I've already mentioned one benefit of the dog genome project, the discovery and location of tens of thousands of points (we called them SNPs) where the genome varies between dogs. They are known as genetic markers, and that is just what they are, molecular flags at intervals along the genome.

The likelihood, in a pedigree dog, is that the hip dysplasia mutation occurred just once and has spread to some of its dog descendants in succeeding generations. Although the initial mutation may well have happened many generations back, the joker (the card causing the disease) will still be flanked by the same genetic markers. Although we can't recognise the joker, we can look to see which markers have travelled with it through the generations. If all or most of the dogs with hip dysplasia also have the same marker (the five of clubs, let's say), then we can assume that the joker (the dysplasia gene) is nearby. We can then look carefully at those genes lying close to the five of clubs in our pack and, by sequencing them, locate not only the dysplasia gene but, if we are lucky, also the actual mutation in that gene which causes the disease. It's a very powerful method and has been used, with minor variations, in identifying the genes for several inherited human diseases. In many cases the mutations were in hitherto unknown genes, so 'guessing' was never going to find them.

There is also an evolutionary history written in recombination. Going back to our metaphorical card decks for a moment, the cutting and shuffling that goes on in every generation will continue to mix the order of the cards. After just one generation, the recombined deck contains one long run of red cards followed by a long run of black. In successive generations the runs of colour will get shorter and shorter as the genomes mix.

This gradual mixing over time can be followed with the same genetic markers that were used to locate the dysplasia gene. It is a useful way of exploring how long a breed has been evolving. The longer that it has been closed to outsiders, as current registration rules insist, the more chromosome shuffling will have taken place. The runs of unmixed cards will get shorter and shorter over time. This means that a recently evolved breed will have longer stretches of chromosomes uninterrupted by recombination than will a long-established breed.

These segments are known as 'haplotype blocks' and their length can be measured. This approach can dissect the genetic structure within breeds even more closely than can the measures of overall genetic similarity method used by Bridgett vonHoldt and Robert Wayne in 2010. Seven years later, in 2017, Heidi Parker published her study of haplotype blocks.[1] Unsurprisingly, the overall relationships between breeds closely resembled those drawn earlier with unsorted genomic data, but added some new twists. For the first time, we could see where two breeds shared haplotype blocks from a common ancestor even though they were not particularly closely related overall. This was delving into the genetic history of breeds not with a magnifying glass but with a microscope.

Closer inspection of the results adds a wealth of detail both about the history of domestication and breed formation in general and the peculiarities of particular breeds. Whereas the earlier unsorted genomic trees gave us a good idea of the overall genetic differences between breeds, the haplotype block approach adds a time element.

As generations pass, recombination slowly breaks up the linked haplotype blocks as the chromosomes are broken and re-formed. The positions of these break points are to all intents and purposes random, so that the haplotype blocks become gradually shorter and shorter over time. This means that by estimating the average length of shared haplotype blocks between two breeds we get a good idea of how long in the past they have been evolving separately.

The African Basenji shares the shortest segments of any with other dog breeds, confirming its reputation as a very unusual dog with a long independent history. Long used for hunting both by sight and scent in many parts of Africa, the Basenji has some distinctly odd habits for a dog. For a start, it does not bark. Instead it chortles and growls and occasionally starts to yodel. Unlike other dogs but in common with wolves, it breeds only once a year. It is very fastidious in its grooming, just like a cat, and also like a cat it is happy to spend all day just staring out of the window. Because of these unusual features and its superficial resemblance to images from ancient Egypt, many people have thought of the Basenji as a very ancient breed and perhaps the one most closely resembling the first domesticated dogs.

The haplotype-sharing genetic analysis certainly marks the Basenji out as a breed with an unusual evolutionary history, more or less separate from all other modern dog breeds. However, this does not necessarily mean, as some think, that the Basenji resembles the first domesticated dogs more closely than any other breed. Rather, it implies that its ancestry has been separate and it has been evolving on its own for a long time.

An interesting illustration of the effect of sorting breeds according to their shared haplotype blocks is found in the Eurasier. This is a breed of whose history we know a great deal. The breed was developed in Germany by Julius Wipfel and Charlotte Baldamus in 1960 for the express purpose of creating the perfect companion dog. Wipfel and Baldamus wanted a dog that showed some original wolfish features, not some pampered breed which had all the wildness bred out of them. They settled on crossing a Chow with a Wolfspitz. The crossing was successful and the Eurasier has since become a favourite companion dog with a devoted following, which included the Austrian animal behaviourist Konrad Lorenz. Lorenz believed that his Eurasier, Babett, had the best character of all the dogs he had known.

The two parent breeds came from entirely different backgrounds, the Chow from East Asia and the Wolfspitz from Germany. Heidi Parker and her team included a Eurasier in their project aimed at assessing the amount of hybridisation between different breeds. She did this by breaking down the dog genome into convenient haplotype blocks each containing a hundred SNP markers spread over at least 232 kilobases (kb). Then she measured the number of these blocks that were shared by pairs of dogs from different breeds. The rationale was that the higher the degree of sharing between the pairs of breeds, the more hybridisation had played a part in the breed's formation. As one might expect, the degree of sharing between breeds of the same functional type (hunting, working etc) was much higher, about four times, than with breeds from different clusters in the original phylogram (Wayne/vonHoldt, page 117). However, when it came to the Eurasier, Parker's analysis found that the breed shared haplotypes with both the two 'parent' breeds, the Keeshund, closely related to the Wolfspitz, and the Chow. Because the origins of the two parent breeds were geographically so different, they had few haplotype blocks in common and the Eurasier formed a new clade (a group evolved from one common ancestor) on its own, lying between the two parent breeds on the resulting phylogram.

Most 'regular' breeds, by which I mean those formed from dogs of similar background, were confined to a single clade with strong statistical support. However, there were some breeds that did not fit neatly into well-defined clades, including the Belgian sheepdog, known as the Tervuren, the Cane Corso, sometimes called the Italian Mastiff, the unmistakable Bull Terrier and its miniature version, the feisty American Rat Terrier, and the American Hairless Terrier, a modern breed created from a bald Darwinian 'sport' picked from a Rat Terrier litter. These breeds lay only just outside the single clade rule by this test, but four other breeds stood out from the crowd. They were the American Redbone Coonhound, the Jack Russell terrier, the Sloughi, a sight-hound from North Africa, and the Cane Paratore, an ancient herding dog from the mountainous Abruzzi region of Italy. These misfits hint at a murky past, perhaps being not quite as pure as aficionados of these breeds might prefer. Other breeds that were adrift were either new breeds 'under development' in the sense that they had not been fully defined, or the same breed collected in different countries. The Cane Corso is a good example of the latter. In Italy, the birthplace of the breed, the Cane Corso forms a discrete clade, whereas in the USA the breed forms a mixed clade related to the, very similar, Neapolitan mastiffs. This all tells us that the genetic structure of these two closely related dog breeds has already begun to differentiate.

Parker and her colleagues took the next step, which was to examine the similarity of breeds, not just from the point of view of measuring the extent of gene-sharing, but looking at the surrounding areas on each chromosome. To explain this I will need to wind back for a moment. I have blithely referred to 'gene-sharing' without being clear what I meant by it. All dogs have the same basic set of genes which carry the information about how to build and run a dog from one generation to the next. Genes are, strictly speaking, bits of DNA that *do* something rather than just being there. Some dog genes control the structure of the eyes, others dictate the form of the bones or the length and colour of the coat. All dogs have these, so in that respect they share 100 per cent of their genes. That makes for a very dull comparison between breeds – they are all the same.

What I am really talking about is the sharing of *different versions* of the *same* gene. For instance, a black dog will have a 'black' version of a coat colour gene and a white dog will have a 'white' version of the same gene. The precise DNA sequence of the two versions will be different, and this difference can be detected in a way I will explain a little later.

I hesitate to introduce a new term because I know from teaching genetics for many years that it is a hard one to grasp. But here goes. The different versions of the same gene are called *alleles*, short for allelomorphs or 'other forms'. When I refer to 'gene-sharing' between breeds, what I really mean is 'allele-sharing'. For the rest of the book I will carry on using the misnomer 'gene-sharing' because this is most definitely not a genetics textbook, there is no exam at the end of term and I don't want you to miss the point of what I am trying to get across by thinking 'What on earth was it that he meant by allele?' and thereby losing the thread.

This is also as good a time as any to say more about how DNA is analysed these days. As you know by now, DNA is literally a long string of four simple chemicals. Very long, actually. It is a code which gives instructions to cells to make this or that protein, like haemoglobin for blood, collagen for bone and all the thousands of other things a cell and a body needs. Like any written language, the sequence of the letters, or in this case the chemicals, is the message. There are only four, abbreviated to A, C, G and T, but even this very restricted selection makes for an infinite number of different sequences if they are long enough. And long they are. The entire dog genome, first sequenced in 2005, is roughly 2.8 billion letters (or 'bases' as a nod to their chemical structure) in total length. That's about the same length as the human genome, but the DNA is spread, as we have already seen, between thirty-nine pairs of chromosomes rather than the twenty-three pairs we humans possess. As we have also seen, there are about 19,000 dog genes, about 3,000 fewer than we humans have.

The daunting task of actually sequencing DNA began in 1977 when the outstanding Cambridge molecular biologist Fred Sanger published the entire sequence of the simple virus called Phi X 174. Its genome, tiny in comparison to the dog or the human, was still 5,375 base pairs (bp) long. Sanger followed this up in 1981 with the first sequence of human mitochondrial DNA, which became the foundation for the whole field of using

DNA to explore the past, something I have been doing for the past twenty-five years. I and thousands of other scientists owe Fred Sanger a great deal, and his contribution to science was recognised by the award of not one but two Nobel Prizes in Chemistry, once in 1958 for the amino-acid sequence of insulin (he bought his supply from the Cambridge branch of Boots the Chemists!) and again, in 1980, for his DNA work. I remember seeing Sanger give a lecture in Oxford around that time and thinking how very self-effacing he seemed. He hated giving lectures and, allegedly, employed a secretary specifically to decline on his behalf the myriad invitations to speak that inevitably follow in the wake of a Nobel Prize. Fortunately for me, he made a rare exception of the Oxford invitation.

Sanger's method is still in use, and I still use it myself for mitochondrial DNA. Both the human genome and dog genome sequences were first completed using versions of the Sanger method automated on an industrial scale. Nevertheless, the former took fifteen years and cost over a billion dollars. The time and expense involved encouraged scientists to think of other, faster, cheaper ways of sequencing DNA and they came up with several ingenious methods. The eventual winner was invented by two French scientists, Bruno Canard and Simon Sarfari, then further developed by two Cambridge chemists, Shankar Balasubramanian and David Klenerman. In contrast to Sanger, and in a sign of the times, they founded a company, called Solexa, to commercialise their methods. Solexa was eventually acquired by the US tech company Illumina and it is they who now supply the sequencers and reagents for what has become known paradoxically as 'next-generation' sequencing.

Without going into too much detail, the Illumina method adds bases attached to fluorescent dyes to a growing DNA strand that is chemically tethered to a small glass flow-cell and copies the exact sequence of the original DNA. As the bases are added one by one they give off a burst of light which is captured by a camera. The colour of the flash depends on which base is added: blue for G, red for T, green for C and yellow for A.

At the end of the reaction cycle there about ninety bases added to each tethered strand. The camera has recorded the order of fluorescent flashes – blue, blue, red, green and so on – in the same sequence that the bases were added, which directly translates to the sequence of the original. There could be millions of strands tethered to the flow-cell, each of them from a different fragment of DNA, and some pretty sophisticated software is required to sort out the flashes and assemble the short fragments into an overall sequence that, on a good day, covers the entire genome.

Thus in the forty years since Fred Sanger painstakingly sequenced the 5,000 or so DNA bases in Phi X 174, the latest Illumina machines can get through fifty complete genomes, each of three gigabases (3 x 109 bases), every day.

19

At the Heart of the Matter

The large body of work done on the dog genome has thrown up many interesting facets of the molecular mechanisms which lie behind the transformation from ancient wolf to modern dog. Unsurprisingly, the greatest successes have been in explaining the physical changes. They are, after all, considerably easier to study.

Modern geneticists live in a digital world where mutations in individual genes lie at the heart of their world view. There was a time, until about ten years ago, when we (and here I am counting myself among their number) dreamed that DNA would explain the whole of biology and medicine. Such youthful arrogance is characteristic of a new discipline and we were certainly not lacking in that department. As time went on, the cracks began to show. The 'gene' for something discovered in one laboratory had mysteriously disappeared when another research team tried to find it. As the geneticist, author and wit Steve Jones once remarked of the frustrations that plagued the gene hunters, the whole fiasco resembled the antics of T. S. Eliot's feline hero Macavity:

> Macavity's a Mystery Cat: he's called the Hidden Paw –
> For he's the master criminal who can defy the Law.
> He's the bafflement of Scotland Yard, the Flying Squad's despair:
> For when they reach the scene of crime – Macavity's not there![1]

In fact it was in an effort to ease the frustration caused by Macavity's unpredictable peregrinations around the human genome that so much money and effort was poured into its canine equivalent, and for this we must be grateful to the cheeky feline. And, my goodness, the human genome is certainly covered in his rascally paw prints.

In a recent paper Jonathan Pritchard and his colleagues at Stanford University reviewed the findings of a large number of searches for genes controlling human height.[2] There were a bewildering 697 locations in the human genome (and by implication at least the same number of genes) that between them accounted for only 16 per cent of the variance. Pritchard and his team searched again for the 'missing' genes, the equivalent of cosmology's dark matter. They were dismayed to find that innocuous common variants, each with tiny effect, accounted for an astonishing 86 per cent of the heritability. They went on to make the sobering estimate that 62 per cent of all common SNPs affect height, very few of them lying in protein-coding regions.

Surely it would be easier to find important genes in the dog. And I am glad to report that so it has proved. This is largely because, unlike humans, dogs are drawn from a limited lupine gene pool in the first place, which has been further reduced by selection and, in pedigree dogs, closed breed formation, as we have seen. This makes gene-hunting a lot easier in dogs than it is in humans where, outside closed communities such as the Old Order Amish, mating is more or less random. In vivid contrast to the hopelessly muddled human situation, characteristics like body size in dogs are indeed controlled by only a few genes that in several cases have been identified. Even though the whole range of size, shape and general appearance is immensely wide in the domestic dog – more so than in any other species – only relatively few genes are involved in this morphological cornucopia. From the sleek and powerful Great Dane to the diminutive Pomeranian, from the deep-pile luxury of the handsome Komodoro to the Chinese Crested, practically amphibian in its nakedness, their varied morphologies are explained by just a few genes.

The first study to discover the genes contributing to overall size used a large collection of Portuguese Water Dogs, 330 in all, recruited through the Georgie project – a not untypical combination of public participation in a science project.[3] Georgie was a Portuguese Water Dog that her owner Gordon Lark had acquired as a stray in 1986. In his own words, he quickly fell in love with her and with her breed, which had been a favourite among Portuguese coastal fishermen. As the name of the breed suggests, they love the water and are superb swimmers both above and below the surface. As a breed they are 'fiercely loyal and with boundless energy', though I must add here that I have never read a breed description that was anything less than flattering. Georgie died in 1996 of an autoimmune disease, and in a series of events reminiscent of the reaction to the tragic death of a child, the project that bears her name was born. Its stated aim was to use modern science to investigate and, one day perhaps, to find a cure for Georgie's illness.

The project came to the notice of Elaine Ostrander, one of the handful of experienced canine geneticists, who chose the Portuguese Water Dog as the subject for her study of the genetic factors involved in skeletal morphology. Having myself been involved with projects requiring large numbers of (human) subjects, I can tell you that having the enthusiastic backing of motivated volunteers is absolutely central to their success. All the dogs were X-rayed and a series of measurements taken which described the morphology of each dog in mathematical terms, which could then be amalgamated into a single numerical description.

Importantly, the Portuguese Water Dogs used by Ostrander came from a limited number of founders made up of only 31 dogs some 24 generations in the past. Using 460 Portuguese Water Dogs and a series of 500 microsatellite markers, the forerunners of SNPs, Ostrander went through the genome segment by segment, looking for those markers that correlated most closely to the skeletal metrics. Knowing the details of each pedigree, she was able to assess the degree to which these metrics were inherited. It is certainly a complicated process but the outcome is relatively straightforward – a map of the genome showing the approximate locations of the genes involved in shaping the skeleton. Whereas as we saw earlier a rather similar analysis in humans by Jonathan Pritchard identified hundreds of places along the human genome that harboured genes controlling height, the Georgie project identified only six of any significance. These are gene locations rather than genes per se, so further work was needed to sift through the regions until the genes themselves were revealed. The forerunner to the dog genome project had already identified a promising gene called IGF1 very close to the most influential segment in the Portuguese Water Dog. IGF1, or, to give it its full name, insulin-like growth factor number one, is a protein involved in the activation of growth hormone and is widely implicated in bone and tissue growth. It is known to be faulty in the human growth disorder Laron dwarfism. However, despite high expectations there were no variations within IGF1 which could explain any effect on growth in the Portuguese Water Dog. More than likely the explanation lay in a genetic change close to the IGF1 gene which somehow modified its behaviour. However, what really excited Ostrander and her team was that the very same genetic segment containing IGF1 identified in the Portuguese Water Dog was present in all small dog breeds and rare or absent in all large breeds.[4]

The conclusion from this unexpected finding was that overall size in dogs was primarily controlled not by hundreds of different genes, as in humans, but by a single gene located very close to IGF1. The fact that it was found on the three different genetic backgrounds in various breeds suggests that either the mutation happened three times independently or is an old one which has spread by what is called a selective sweep into all miniature and toy breeds. There could not be a more dramatic demonstration of the huge morphological change introduced by a single gene.

appearance of visibly enhanced muscles in a litter of pups. These pups grew up to become excessively heavy and strong and thus unsuitable for racing. They are usually put down. However, breeders also noticed that the parents of the Bully Whippets and some of the littermates were slightly more muscular than normal and also faster on the racetrack. This combination of mixed litters of Bully Whippets, partial 'bullies' and normal pups is typical of what is called a recessive genetic condition. The full 'bullies' had inherited two copies of the 'bully' gene, whereas the partial bullies received only one and the normal pups none. This pattern of inheritance is reproduced in all recessive genetic conditions, including the well-known human disorder cystic fibrosis.

Both the Bully Whippet and patients with cystic fibrosis and other recessively-inherited disorders are at a significant breeding disadvantage, the dogs because they are put down and the humans because they are likely to be incapacitated before they have children. This raises the very interesting question of why these conditions are as common as they are, since the genes responsible are being continuously eliminated by selection every time an affected individual, dog or human, fails to breed. In fact, not one but two copies of the gene are lost from the population as a whole when these 'double dose' individuals, or homozygotes to use the scientific term, die without leaving offspring.

The answer is an elegant and important one for understanding how genetic conditions become established. Although individuals, dogs or humans, with double doses of a mutant gene, that is the homozygotes, have fewer offspring, some of their siblings, or litter mates, will have one mutant and one normal gene. If these carriers, called heterozygotes, possess some advantage over individuals carrying two copies of the normal gene, they will prosper. In the case of Bully Whippets, carriers can run faster than regular dogs and will be chosen for breeding. The same rule applies to humans. Carriers of the cystic fibrosis gene must have had an advantage, not necessarily now but in the past, and the most likely is an increased resistance to cholera. In a thoroughly researched tropical disease called thalassaemia, which is very common in the Mediterranean and in south-east Asia, it is the advantage conferred on carriers in being able to resist infection by malaria that gives them the edge. The disease persists because, even long after malaria has been eliminated, as it has in the Mediterranean, the genes are still there.

The genetic explanation for the Bully Whippet is equally interesting and owes its solution, like many genetic mysteries, to the observation of a single human patient by a keen-eyed doctor. In this case the location for the breakthrough was the paediatric department of the University Medical Centre in Berlin.[8] After a normal pregnancy, a healthy woman gave birth to a son. Soon after birth the boy developed sudden involuntary muscular contractions, called myoclonus, and was admitted to the neonatal ward for observation. He appeared to be extraordinarily muscular, with bulging thigh and arm muscles, but was otherwise completely normal. The myoclonus subsided after a couple of months and the boy continued to develop normally though he remained extraordinarily muscular. His case report and genetic investigation were written up when he was four and a half, by which age he was able to lift two 3-kilogram dumbbells with his arms straight out.

Research into the cause of his condition chose the candidate gene approach with the myostatin gene as its focus. Myostatin itself is a protein which, like IGF1, is a growth factor, but instead of being involved in bone growth its effects are on muscle development. The choice of myostatin as a candidate was a good one, and that was indeed the location of the mutated gene.[9] However, the mutation itself, far from leading to an increased level of myostatin in the German boy, actually cancelled it altogether. It turned out that myostatin slows down muscle development, so when there is no myostatin its dampening effect is removed and muscle growth proceeds unchecked. This is a common type of mutation, called a 'null', giving the boy a homozygous null/null genotype. As it's a recessive condition, both of his parents must have been carriers with null/wild type (*wild type* being the geneticist's label for normal) genotypes, which, as we have seen, probably confers some advantage. Indeed, the boy's mother was in her youth an Olympic-class swimmer and many of her relatives were unusually strong. This raises the interesting ethical point, entirely incidental to our consideration of the Bully Whippet, as to whether carriers of this condition have an unfair advantage in competition. The general consensus is that though the partial inactivation of myostatin might confer an advantage, it is but one of the multitude of genetic influences which elevate elite athletes above the rest and should not be discriminated against.

The myostatin mutation in the Bully Whippet, as with the young boy from Berlin, was one that inactivated the gene, but it did so in an interesting way. Most mammalian genes are separated into two classes of DNA sequence. The first, called the exons, contain the DNA sequence which directly encodes the amino acid sequence of the gene product, in this case the myostatin protein. Any mutation in the coding regions of these exons risks changing the amino acid sequence of the protein with potentially disastrous consequences. On sequencing the myostatin gene in twenty-two whippets, researchers discovered that all four 'Bullies' were missing a small DNA segment only two base pairs in length. This was enough to completely inactivate the gene because cells read genetic instructions in groups of three. The sequence determines not just the order of amino acids in the final protein but also their identity.

The two-base deletion causes what is called a frameshift mutation. The cell doesn't know whether the DNA sequence it is reading is the correct one or not, it just carries on adding amino acids one by one in linear order according to the instructions it receives from the DNA sequence. The protein strand grows from one end and terminates at the other. In the Bully Whippet the myostatin strand, following instructions from the myostatin gene, grows normally for the first few hundred amino acids. Then it encounters the deletion, which causes a frameshift. Because the sequence is read in groups of three nucleotides, the amino acid sequence on the downstream side of the mutation is completely jumbled. Worse still, the mutation converts the normal three-base signal for the amino acid cysteine into one which promptly terminates the synthesis of the protein and destroys its function. All this because of a two-base deletion in a genome hundreds of millions of bases long. That is all it takes.

Strictly irrelevant in a book about dogs but nonetheless interesting is a historical observation dating back to 1807 of a similar double muscling in Belgian Blue cattle, which turns out to be due to a very similar mutation in the myostatin gene.[10] Unlike whippet breeders, concerned as they are with speed, where the muscular Bully Whippets are too bulky to compete and are disposed of, breeders of beef cattle have been very keen to select for the bulky bovines. Although breeders knew nothing of the molecular genetics at the heart of the issue, herds of Belgian Blues and the closely related Piedmontese breeds have been built up by carefully controlled breeding so that *all* the animals now carry the myostatin mutation in double dose.

In situations like this where all the animals carry the mutation, it is said to be 'fixed'. Fixation of the double muscling myostatin mutation is complete in those herds of Belgian Blues because breeders have intentionally selected it. In the more usual situation the population as a whole contains mainly wild-type individuals with a smattering of mutants whose future career depends on how fast they are eliminated by selection. Without a compensating carrier advantage, damaging gene mutations tend not to last for more than a few generations.

There is a well-known example of a mutation that has become fixed in one dog breed, the Dalmatian. All Dalmatians, not just some, suffer from very high serum levels of uric acid. In humans high serum uric acid leads to gout when the uric acid forms crystals in the kidney and in the joints, particularly the big toe. Uric acid is the final waste product from the breakdown of purines, one of the chemical components of DNA, but hyperuraemia, as a high blood level of uric acid is called, is only a problem in humans, great apes like gorillas and chimpanzees – and Dalmatians. All other species and all other dog breeds break down purines in a different and less troublesome fashion.

The human gene involved in gout had already been identified, so the first step to tracking down the Dalmatian mutation was to sequence the gene in the dogs. However, this led nowhere and the search for the gene had to begin all over again. After a great deal of work involving crosses between Dalmatians and Pointers, the gene was located to an area of the genome containing just four genes. Detailed sequencing of these genes revealed that the causal mutation was a single DNA base change in a gene called SLC2A9, which encodes a protein that helps glucose and uric acid cross cell membranes.[11] Unlike the deletion we saw in the Bully Whippet, this change did not alter the reading frame but it did switch the amino acid sequence at that point, substituting a phenylalanine for a cysteine, which is enough to inactivate the gene. Although Dalmatians are the only breed in which all dogs suffer from hyperuraemia, a survey of other dogs which had been diagnosed with the disease showed that in two breeds, Bulldogs and Russian Black Terriers, some dogs carried precisely the same mutation. This must mean that the mutant gene had originated before Dalmatians became genetically isolated as a breed.

An interesting question is why the 'gout' mutation is fixed in the breed. Why are all Dalmatians homozygotes for the mutant gene, while it is very scarce in other breeds? The answer lies in a nearby gene which enhances the appearance of the characteristic spots on the dog's coat, a genetic trait that appealed to breeders and led to selection not just for this benign coat-pattern but also for its far from harmless neighbour. Being so close to each other on the same chromosome, these two genes are inherited together every time.

In the Dalmatian, as in most dog breeds, appearance matters a lot. It's unfortunate that the characteristic and highly desirable spotting pattern in the breed brings with it a serious disease just because of the proximity of the piebald coat-pattern and purine metabolism genes. Selection for one was inevitably linked to selection of the other, albeit unintentionally.

Appearance has always been important for dog breeders, as compliance to rigid breed standards is so influential in judging competitive shows. Important among these, perhaps the most important, are the characteristics of the coat – its colour, thickness, curliness and so on. It was natural therefore for the architects of the dog genome project to try and find the major genes responsible for these important features. Such is the variety of patterns and textures that one might be forgiven for thinking that dozens or even hundreds of genes might be involved. But one of the hopes of the dog genome project was that, in comparison to humans, apparently complex situations might be explained by just a few genes.

To test this important principle the team at the Broad Institute at Harvard that sequenced the dog genome threw their whole impressive technical armoury at the project. They chose to focus the search on two features already known to have a conventional form of inheritance – the 'ridgeless' form of the Rhodesian Ridgeback and white coat-colour in Boxers. Both traits are simple recessives, meaning that both parents have to be carriers. However, it is clear that this technical *tour de force* had wider ambitions than locating two quirky genes in two minor breeds (if Rhodesian Ridgeback and Boxer owners will forgive the slight). The Broad team was clearly trying to formulate a general approach to finding interesting dog genes, and these two were working examples to demonstrate the success of their methods. In the genetic equivalent of saturation bombing, they selected 27,000 equally spaced SNPs across the canine genome, then typed 250 dogs of various breeds and measured the average length of DNA haplotype blocks inherited through the generations. They found that haplotype blocks within breeds are generally long and contain only a handful of common haplotypes. This they interpreted as a reflection of the population history of the breed.

For example, although the Shiba Inu breed is an ancient one, its numbers were drastically reduced by the Second World War and it was nearly wiped out. A few dogs survived, and all today's Shiba Inu are descended from these lucky few – and are all very closely related. Consequently, the average length of haplotype blocks in the Shiba Inu is long, the longest of any breed. In contrast, the Greyhound, also an ancient breed, has not suffered from a catastrophic population decline. The Greyhound has maintained a large breeding population and has the shortest average haplotype block length, abbreviated to LD, of any breed. Even so, the LD of all pedigree breeds is much longer than in dogs in general, and in feral dogs especially where mating is completely random. The longer LD in pedigree breeds is explained by the relatively small number of founder animals for any breed involved in its creation. LD values for dogs in general are much higher than the equivalent wild populations of wolves from which they descend. This level of LD is a reflection of the relatively small number of wolves involved in the creation of 'domesticated' dogs in the first place, followed by thousands of years of selection.

It did not take long for the Broad's saturation mapping campaign to identify the regions of the genome wherein resided the two targeted genes. In the Rhodesian Ridgeback, the dorsal ridge of inverted hair which is the characteristic of the breed was soon located to a 750 kb region of chromosome 18 containing five known genes, three of which are growth factors involved in embryonic development.[12] Having located the genes, the next step, rapidly accomplished, was to sequence them and find out what was going on in the Ridgeback. The answer turned out to be a different type of mutation than we have covered in the Bully Whippet and the Dalmatian. There the mutations had deleted short segments of DNA. In the Rhodesian Ridgeback the opposite had happened. A large chunk of DNA containing all three growth factor genes had been duplicated. As you will begin to appreciate, the genome can get up to all sorts of tricks. The effect of the duplication was to give the dogs a double helping of these growth factors which did them no good, other than creating the dorsal ridge which defines the breed. In the uncommon 'ridgeless' variant there was no duplication and just the normal arrangement of these growth factor genes, just like other breeds. So the ridge on the Ridgeback is a consequence of deliberate selection for the feature. Unfortunately, the 'ridgeback' mutation also predisposes dogs to neural tube defects similar to dermal sinuses in humans, a similar type of defect to spina bifida. Once again, as in the Dalmatian, selection for one characteristic feature also brings with it a less welcome companion.

When the Broad team turned its attention to the Boxer they were looking for the gene responsible for the absence of skin and coat pigmentation. Once again this is a recessively inherited trait, with carrier parents having an intermediate colouration with patches of white on a solid background. Earlier breeding experiments in the 1950s had suggested that the same genetic flaw was behind so-called 'Irish spotting' in the Basenji and Bernese Mountain dogs and piebald coats in Beagles, Fox Terriers and English Springer Spaniels.

The saturation mapping technique that had proven so successful with the Rhodesian Ridgeback soon narrowed the Boxer pigmentation gene to a one megabase (1 million base) segment that contained just one gene.[13] This gene (MITF) is important in embryonic development and is mainly concerned with the production of melanocytes, the cells which produce the pigment melanin.

In the Boxer, it proved difficult to narrow down the mutation to a particular set of DNA bases because there was not enough genetic variation. This made it impossible to distinguish mutations which caused the pigmentation and others that were just normal variations. Lack of genetic variation is the flip side of hunting for causal mutations in closed pedigree breeds. In order to overcome this issue the team used a second species, the Bull Terrier, where the genetic variation was greater and which has a similar white version. Even so, they could not be completely confident of identifying the DNA change which caused the white pigmentation. This too is not an uncommon result, even with the impressive array of genetic tools recruited to the task by the Broad Institute, but they were eventually able to locate it to MITF encoding a cell transcription factor, a family of proteins involved with turning particular genes on and off.

The same gene (MITF) is also implicated in Waardenburg syndrome in humans. Here it is associated with deafness, cleft lip, a white streak in the hair and sometimes brilliantly blue eyes or even eyes of different colour – or heterochromia. The arresting visual appearance of heterochromia lends an alien quality to those who have it, notably David Bowie in his incarnation as Ziggy Stardust. And who can forget the arresting face of the young Afghan girl with piercing blue eyes who reminded us all of the human cost of that terrible war? Genetics makes the most unexpected connections.

A couple of years later the veteran duo of Robert Wayne and Elaine Ostrander published a wide-ranging genetic study on coat variation which came to the remarkable conclusion that only three genes were between them responsible for the dizzying array seen in domestic dogs.[14] Following the successful and by now familiar saturation mapping used by the Broad Institute team, Wayne and Ostrander concentrated on three coat characteristics. They chose first the presence or absence of 'furnishings', the name given to the eyebrows and moustache typical of wirehaired dogs such as the Highland Terrier. Second was hair length, and finally a third characteristic – whether the hair was straight or curly. They first typed the SNP panel in 96 Dachshunds with three different coat-colour varieties – wirehaired with furnishings, smooth, and finally long-haired. This located the furnishings gene to a segment of chromosome 13 that contained just one gene – RSPO2, an excellent candidate that had already been implicated in the development of hair follicle tumours which occur predominantly in breeds that have furnishings. As with some other canine genetic features there is a human equivalent – an East Asian hair type with some resemblance to wirehair in dogs. This is not caused by the same gene but by one called EDAR, which is also involved in the same developmental hair growth pathway.

The furnishings mutation turned out on detailed analysis to be yet another type that we have not yet encountered. There was nothing abnormal about the RSPO2 exons; the sequences were all entirely normal. However, there was an insertion just outside the gene itself which, it turned out, modified its expression, increasing it threefold and presumably leading to the hirsute character of the dogs that possessed it.

The team then turned their attention to fluffiness in Welsh Corgis, a breed in which dogs can be separated into long-haired, and therefore fluffy, and shorthaired non-fluffy individuals. Typing both varieties by saturation mapping pinpointed the mutation to one of the growth factor genes, FGFR5. This time the mutation was a familiar one, a straightforward substitution of amino acid, phenylalanine, for another, cysteine. You may have noticed that cysteine features a lot in the mutations we have illustrated. This is because cysteine forms connecting molecular bridges between protein chains and, more often than not, these connections are essential for making sure the protein chains come together in the correct orientation.

In order to investigate curliness the team chose that old favourite, the Portuguese Water Dog, which occurs in two forms, curly haired and wavy haired. The same mapping technique soon found that the gene responsible for these two hair types was a member of the keratin family and the mutation was a straightforward substitution of a single amino acid arginine for another, tryptophan, caused by a single base change.

What makes this energetic survey especially noteworthy is that by studying combinations of only three genes they had identified pretty well all variation in hair type seen in all breeds. For example, Basset Hounds have wild-type alleles at all three genes, resulting in a shorthaired non-curly coat without furnishings. Wirehaired Australian Terriers possess the wild-type variants (Wt) at FGFR5 and KRT71 but the mutant form of RSPO2 which confers a wired coat form. To avoid unnecessary repetition, I have summarised the outcome in the table below.

Type	Example	FGFR5	RSPO2	KRT71
Short	Basset Hound	Wt	Wt	Wt
Wire	Australian Terrier	Wt	Mutant	Wt
Wire & Curly	Airedale Terrier	Wt	Mutant	Mutant
Long	Golden Retriever	Mutant	Wt	Wt
Long with furnishings	Bearded Collie	Mutant	Mutant	Wt
Curly	Irish Water Spaniel	Mutant	Wt	Mutant
Curly with furnishings	Bichon-Frise	Mutant	Mutant	Mutant

None of the three mutant genes was found in three grey wolves, an admittedly tiny sample, nor in any of the shorthaired dogs. This strongly suggests that the ancestral type was shorthaired, without furnishings and not curly. The implication of this is that all breeds with any of these derived features started off at some stage in the past with the ancestral type and that the mutant forms were introduced at a later stage during breed formation.

The saturation mapping that was so successful in localising these three genes also showed that the haplotypes surrounding each of them was exactly the same in all breeds. This means that, rather like the IGF1 gene for size, these three coat-type genes were each initially confined to one breed, even one litter, and then, through selective mating, spread to all the others.

Finally in this consideration of the effects of mutations I will turn to colour. In most mammals natural pigmentation is controlled by the melanocortin 1 receptor pathway. This pathway moderates the type and amount of melanin pigment made in the skin and hair. The huge range of colour that we see in the animal kingdom is due to just two forms of melanin: the red/yellow pigment pheomelanin, and the brown/black eumelanin. The subtle range of colouration is due to the way melanocytes moderate this very restricted palette. There are a few other genes involved, notably within what is called the 'K locus', which produces a protein capable of modifying the melanocortin receptor pathway and was found to be crucial in very black dogs. Melanistic dogs are unsurprisingly derived, presumably by selection, from the wild-type ancestor. However, the study that brought this to light came up with a major shock when it included melanistic wolves. Black wolves are only found in North America, except for a few in Italy.

In an effort to locate the melanistic gene, or genes, in North American wolves, the research team, which included the omnipresent Robert Wayne, made arrangements to study the Leopold pack, one of the wolf packs to form after the reintroduction of Canadian wolves into Yellowstone National Park in 1995. The subsequent activities and breeding structure of the pack have been intensively studied and a precise genealogy constructed. The Leopold pack contains both black and wild-type individuals, so it was comparatively straightforward (other than catching them, obviously) to see whether MCR1 or the K locus was inherited along with the colouration. This study showed a perfect split between genetic variants at the K locus and whether or not wolves were black. So far so good.

The surprise came with the genetic analysis of the same region in melanistic domestic dogs. This showed beyond doubt that the mutation, a three-base pair deletion, and adjacent sequences were exactly the same in both wolves and dogs. This had to mean that, contrary to the usual

Resident Yellowstone wolf pack keeping American Elk in the river. Note the colour variation among the wolves.

flow of genes from the wolf to the dog, the K locus had gone the other way both in North America and, it transpired, also in Italy. The black wolves were black because they had inherited the melanistic mutation from domestic dogs. This gene, which has spread into the wild wolf population since domestic dogs reached North America from Asia along with the first humans some 12,000 to 15,000 years ago, is itself under the influence of selection. However, the selection among wild wolves is not artificial but natural, as a survey in the Canadian Arctic showed. In the far north and east where the barren tundra predominates, black wolves are rare, presumably because they are at a disadvantage during the winter when they stand out against the snow, making it harder for them to track prey. However, further west and south the proportion of black wolves increases as tundra gives way to forest, where the dark colouration is a helpful camouflage. It has been known for a long time, and Darwin himself noted it, that Native Americans encouraged the occasional mating of their own dogs with wild wolves in order to introduce a fresh dose of the wild into their dogs. They would tether females in heat to trees and leave the rest to nature. There would need to be some modifications to transfer dog genes to wolves, but somehow that is what happened.

So far we have only considered genetic traits in the dog that affect its physical aspects – size, coat-colour, musculature and so on. These have a rich variety of causes involving a range of different mutation types in a large number of genes. The genetic complexity of these changes, great as they are, have been solved thanks to our recently acquired detailed knowledge of the dog genome. Through this, researchers have identified candidate genes which, given a bit of luck, have often turned up trumps and have been shown to be the location of causal mutations. However, the creation of a set of equally spaced genetic markers covering the entire genome has led to the ability to locate genes when inspired guesswork fails. These saturation mapping techniques make no assumptions about the nature of the genes involved and just plough through the genome until their locations are revealed.

Once those are mapped, like an X marked on a treasure chart, detailed sequencing around the location soon reveals the gene itself and usually the exact mutation behind the trait. The modern structure of pedigree breeds, which have become increasingly closed genetic systems, helps a great deal in these gene mapping enterprises, although, as we have seen, the accompanying genetic isolation promotes inbreeding and the appearance of often damaging recessive traits. All the recent successes depend not only on the extravagant technology but also on a basic requirement of the genetics. To map a gene you need a feature which varies within the breed and which is relatively straightforward to score. For example, the experiments with coat-colour mapped the melanism gene because it was easy to tell which dogs were black and which were not. The trait is said to 'segregate' within the breed. If a trait does not segregate and all dogs are the same, then the genes responsible cannot be mapped however many markers are used. Candidate genes can still be scrutinised for obviously disabling mutations like frameshift deletions, but, as we have seen, this is very much a hit-or-miss affair.

The great successes over the last decade or so since the dog genome was published in 2005 have been in identifying genes responsible for physical genetic traits like coat-colour or inherited disorders. Of at least equal and perhaps greater interest is the genetics which lies behind differences not in appearance but in behaviour. These might be between breeds or between different individuals within the same breed. For example, a bloodhound makes a good hunting dog but is hopeless at herding sheep. Equally an Old English Sheepdog will control a flock of sheep with ease but would be next to useless following a scent through thick undergrowth. These differences between breeds have been known for millennia and indeed have been under intense selection for almost as long as dogs have been our companions. They have a genetic basis, but it has proved extraordinarily difficult to identify the genes involved, for a number of reasons.

First of all, whether or not the trait segregates is an essential requirement for mapping. Certainly, some individual bloodhounds will be better than others at following a scent, and some sheepdogs will be better than others at herding. But unlike coat-colour, where all sorts of variation is tolerated, bloodhounds that have no sense of smell and sheepdogs who just don't get it when it comes to herding are soon dispatched or at the very least not chosen for breeding. This severely limits the range of variation within a breed for geneticists to work with.

Next comes the issue of assessment. It is easy to score a dog as black or white, but how do you measure herding ability, for example, in a reproducible way? Various approaches have been tried, from expert assessment (very slow and expensive) to owner-directed surveys of performance (difficult to achieve consistency). These two very practical considerations have held back advances in understanding the genetics behind these most interesting behavioural differences between breeds. Nonetheless the good old-fashioned candidate gene approach has met with some limited success.

The candidate gene approach succeeded in unravelling the cause of the sleep disorder narcolepsy in Dobermanns.[15] Dogs with this condition tend to fall asleep at a moment's notice, presumably not an asset in a breed created primarily for guard duty and the personal protection of the breed's creator, German tax collector Louis Dobermann. Nevertheless a colony of Dobermanns exists that segregates for the condition, and this allowed researchers to test a number of candidate genes to see if any of them follow the same route through the pedigrees. One of them did and it was identified as the hypocretin receptor gene, where an insertion had disrupted its normal functioning in the brain.[16]

This is a rare example of successful gene hunting in a behavioural trait, but the genetic basis for most dog behaviours, though doubtless genetic in origin, is still shrouded in mystery. Unfortunately this includes most behaviours of interest or even concern to owners. For example, a few breeds, particularly the Bull Terrier, suffer from a disorder which is very reminiscent of obsessive compulsive disorder (OCD) in humans. It is easy to score, as affected dogs chase their own tails. Thus far the gene has not been identified, although the successful treatment of the condition with serotonin re-uptake inhibitors such as clomipramine, which is also used to treat OCD in humans, suggests there may be a common genetic pathway in both species.

Aggression is another behaviour of considerable interest to owners and breeders alike, but studying it comes with its own issues. Many in the dog fanciers' community argue that there is no such thing as a bad dog, only a bad owner, and that no breed should be penalised because of inadequate or careless owners. Nevertheless some breeds have been outlawed, notably the American Pitbull. Undoubtedly this breed, until recently, and probably even now, bred for illegal dog-fighting, has been selected for its performance in the fight-pit. But that does not mean that Pitbulls are intrinsically vicious, as might be indicated by a consistent genetic mutation. One reason I love genetics is the multitude of ethical questions it raises. The issue of the genetic components of aggression in the Pitbull is immediately applicable to our own species. How far does individual responsibility for violent actions extend if it turns out that the cause is a mutation in the genome? That question puzzles and will go on puzzling ethicists and lawmakers for years to come.

I want to end the chapter with a remarkable example recently revealed which may even hold the secret of dog domestication, for so long the holy grail of researchers. It is also a gripping scientific detective story and shows how a single unusual event, observed by lively minds, can lead to amazing revelations. It began in 1961 in Auckland, New Zealand, at Green Lane Hospital, an offshoot of the city's main infirmary. Three cardiologists led by Dr J. C. P. Williams had noticed four young patients suffering from supra-valvular aortic stenosis. The principal feature of this condition is a narrowing of the aorta, the main blood vessel leading from the heart. Not surprisingly, it can have serious or even fatal consequences. The doctors noticed that as well as the cardiovascular problems these four children shared other features including various degrees of mental retardation and unusual faces. These were referred to as 'elfin', owing to slightly pointed ears and crowded teeth. They all had a wide philtrum, the gap between the upper lip and the bottom of the nose, the overall effect making them easily recognisable. Most remarkable of all, these children had extraordinarily engaging characters. They were always smiling, always cheerful and very easy to approach. Where there are disorders that display a variety of symptoms, these are referred to as syndromes and are often given the name of the principal investigator, so this was named after Dr Williams.[17]

Other cases conforming to this syndrome were soon noticed in cardiology clinics around the world. It seemed likely that the aortic stenosis in Williams syndrome might have something to do with the structure of the aortic wall. Like all arteries the aorta is like a thick, elastic tube which expands and contracts to even out the flow from the left ventricle as the heart pumps blood around the body. The component responsible for the elasticity of the aortic wall is the protein elastin. I worked on elastin for my PhD many years ago and it is a remarkable protein which, as the name implies, really is elastic. It is found not just in blood vessel walls but also in the skin, where thin fibres of elastin keep it taut – at least when you're young. As time goes by the connections that join the elastin molecules together break down, and this is accelerated by the action of sunlight – to, on the one hand, the despair of sun-worshippers and, on the other, the delight of manufacturers offering products that claim to halt or even reverse the ageing process.

Soon after the elastin gene was located on human chromosome 7, scientists checked to see if there were any obvious elastin gene mutations in other conditions, like Williams syndrome, where it might be implicated. This soon led to the discovery that the gene resided in one of those parts of the genome which tended to get muddled by losing whole chunks of DNA. The genome is not as stable as we might imagine. In Williams syndrome patients, deletions did indeed take out the elastin gene and some of the genes on either side of it. Large gene deletions like this usually come out of the blue, so there are often sporadic cases with no family history. The sporadic cases also showed similar large-scale deletions involving the elastin gene. This meant that Williams syndrome patients, both familial and sporadic, had only one working elastin gene when they should have had two. This wouldn't usually matter where an enzyme defect is involved, which is why carriers of most recessive disorders are symptomless, getting by perfectly well on just one copy. Where the protein forms part of a structure, like the arterial wall, producing only half as much as normal has consequences. In all cases where one of the elastin genes is deleted, the result is aortic stenosis.

Following this discovery in Williams syndrome, some other patients with aortic stenosis were also found to be missing one of the elastin genes. But they didn't have the other features of Williams syndrome, including the remarkably open personality. Why not? The most likely explanation is that there was another gene lying close by which is missing or mutated in the Williams syndrome patients but not in patients with aortic stenosis alone. What, I hear you ask, has this to do with dogs?

Once again, we are indebted to Elaine Ostrander from the US National Institutes of Health and Bridgett vonHoldt from Princeton and a mixed team of research scientists in different disciplines drawn from a number of other US universities for making the connection. As we know, Dr Ostrander has a long-standing interest in the domestication of dogs. She and the team wondered whether the exaggerated sociability of Williams syndrome patients may also be a feature of domesticated dogs that was absent in their wild ancestor the wolf.

The first hint came from a saturation mapping project of 701 dogs from 85 breeds and 92 Arctic wolves.[18] The question asked was straightforward. Were there any places in the genome that were substantially different in dogs and in wolves? Two regions stood out. The region of greatest difference between the two lay within a gene called SLC24A4, which is mainly concerned with hair and tooth construction. Not far behind in the rankings was another gene called WBSCR17, which lies very close to the elastin gene on dog chromosome 6. There is a remarkable consistency in gene location and order in mammalian genomes, which hinted that this gene might also be responsible for the other features of Williams syndrome in humans. By then, it was known that the Williams syndrome deletion was about 1.5 megabases long and took out a segment containing roughly twenty-eight genes, one of these being WBSCR17.

There followed a series of behavioural tests of the same sort used to diagnose Williams syndrome in humans on eighteen dogs and twelve captive wolves who were accustomed to being with humans. One was puzzle-solving, another measured hypersociability, and the third assessed social interest in strangers. Without going into details of all these tests, let me summarise the results. In the puzzle-solving test, dogs spent most of their time looking at the humans for cues while the wolves just got on with it and achieved higher scores as a result. Next the researchers measured sociability. In the first test the researcher sat quietly, ignoring the subject but looking at the floor on which was drawn a circle one metre in circumference. In a variation of this, the researcher called the animal by name (remember that the wolves were already socialised to humans) and encouraged them to approach. In this test dogs spent far more time within the circle than the wolves, both when the experimenter was known to them and, slightly less often, when they were strangers.

Though the number of animals in the experiment was small, it was enough to convince the scientists that, on the one hand, dogs but not wolves showed the hypersociabilty associated with the Williams syndrome but lacked the ability to act without human guidance. When it came to pinning down the mutations in the dog that might be responsible, there were mixed results. There was plenty of genetic variation in the region, with mutations present in members of the transcription factor II genes. No one is yet quite certain how these genes work or what effect the mutations might have, but I suspect the answer won't be long in coming. What is clear is that this suite of genes is under strong positive selection favouring those animals showing Williams syndrome-like hypersociability above those that do not.

There is a lot more to be done, but this fascinating piece of research, stemming from the observations of three curious New Zealand doctors, has provided the first solid clue that 'domestication' has a significant genetic basis and was not just a consequence of behavioural adaptation.

The research culminating in the fascinating potential link between an obscure human disorder and a genetic basis for domestication involved some of the most sophisticated technology ever applied to the dog genome and has yet to reveal all its secrets. Had artificial selection applied by man picked out genetic traits that made dogs on the one hand friendly towards us and yet on the other unable to think for themselves?

In the Lab

Throughout this book we have referred to the fascinating genetic research carried out on dogs and wolves. Much of it has been done by scientists eager to use dogs as a proxy for humans in medical research. That was the primary drive behind the Dog Genome project. Other scientists like Robert Wayne and Elaine Ostrander have had a long-standing interest in dogs and their evolution from wolves. They have both contributed to the genome project, but as the supergroups, like that from the Broad Institute, have turned their attention to the next project that deserves the application of the dazzling array of technological weaponry at the Institute's disposal, the dog enthusiasts among them have been busy applying the genome results to projects that benefit dogs rather than humans. One such research group, headed by Dr Cathryn Mellersh, is based at the Animal Health Trust (AHT) Laboratories near Newmarket in Cambridgeshire, and she kindly agreed to my visit. Dr Mellersh had completed her PhD at Leicester University. She then spent a few years in Seattle with Dr Ostrander before coming back to the UK to work at the Trust.

The Suffolk town of Newmarket is the epicentre of British flat racing and home to the National Stud, the headquarters of the UK thoroughbred breeding industry. The Stud was built up by William Paul Walker, the son of a wealthy brewer and horse breeder who was concerned at the

UK's shortage of thoroughbred stallions to re-supply the cavalry regiments. Beginning with Walker's bloodstock in 1915, the National Stud has grown to become the centre for supplying a comprehensive range of services to the thoroughbred horse-breeding industry. The Stud has been owned since 2008 by the Jockey Club, the equine equivalent of the Kennel Club. It was the natural home for the Animal Health Trust, a veterinary charity founded in 1942 and devoted to cats, dogs and, of course, horses.

The Trust is located in the spacious grounds of a leafy estate on the outskirts of Newmarket. On my way to the laboratories I passed horses both in spacious paddocks and being led around the grounds. There were dogs too, playing on the greensward under the watchful eye of their handlers. The whole place exuded a feeling of prosperous calm.

It was a pleasure for me to be talking with Dr Cathryn Mellersh, head of the AHT laboratory, about the nitty-gritty of laboratory genetics, which I had always loved. Lazing on the floor of her office were her two pet dogs Libby and Tess who, she told me, she had found in a rescue centre. She wasn't sure but she thinks they might have belonged to some travellers and used for coursing until they became too old. It's not uncommon for dogs that are no longer needed to be tied to a fence or a lamppost and abandoned. It was immediately clear to me that I was in the presence not only of a scientist but also a dedicated 'dog person'. Cathryn had a deep desire to help dogs, not just to use them as a tool for medical research.

The primary focus of research at the AHT is to tackle the health issues caused by inbreeding, the perennial problem of pedigree dogs. As we've covered in earlier chapters, closed breeds are vulnerable to recessively inherited diseases because whereas carriers are usually entirely free of symptoms, the homozygous animals are affected.

In 2008, an investigative documentary called *Pedigree Dogs Exposed* was shown on BBC1, the main terrestrial channel in the UK. This documentary claimed that the inbreeding inherent in maintaining breed standards of pedigree dogs, especially in the show ring, had adversely affected the welfare of several breeds. Screening of the film caused a public outcry and precipitated a crisis for the Kennel Club amid a deluge of negative publicity. Commercial sponsors withdrew their support and the BBC seriously considered terminating coverage of the annual Crufts dog show, the jewel in the crown of the Kennel Club. It was essential for the Kennel Club to show the world that they were well aware of the problems of inbreeding and that they were doing something about it. Soon after this embarrassing exposé, Kennel Club support for genetics at AHC increased substantially, allowing Dr Mellersh and her team to expand their research programme into inherited canine diseases. I do not want to dwell on *Pedigree Dogs Exposed* here. Enough has been said and written elsewhere.

The ultimate ambition of Cathryn's research is to develop diagnostic DNA tests for recessive disorders and use these to eliminate the mutant genes from pedigree breeds.[1] However, as she explained, it isn't all that simple. Even getting a DNA sample is not straightforward. For example, vets in the UK are not allowed to take blood samples just for research, even if the owner consents. Also these pets belong to people and families and the right approach is needed to secure their agreement. All this takes time, and if a blood sample is needed as a source of DNA it cannot be taken until the dog is having other blood tests, at which time a few extra drops can be saved for research use. This is a difficulty I hadn't appreciated at all before I visited Dr Mellersh. As I can now see, it poses a formidable obstacle.

As we touched on in earlier chapters, there are different ways of finding mutations. These days the standard route is through saturation mapping with a panel of SNPs to find which of them segregates with the disease. This requires a group of at least six unrelated dogs who suffer from the disorder under investigation. For some rare diseases in rare breeds, it is difficult to find enough owners who are prepared to allow blood to be taken from their dogs. To sidestep this issue, Dr Mellersh now sequences the entire genome of individual affected dogs using DNA collected with a mouth swab, though this approach too has brought its own problems.

Dogs have a lot of bacteria in their mouths, bacteria with their own DNA. If there's too much of this contamination, it can interfere with the selectivity of the sequencing reactions and you end up with DNA sequences most of which come from the bacteria and not the dog. To combat this difficulty Cathryn's lab does a preliminary sequence on the swab DNA and rejects any samples that are less than 90 per cent dog. The bacterial sequences still get read but can be weeded out during the subsequent computational manoeuvres.

As you can see, in laboratory science, solving practical issues such as these can make the difference between success and failure. The students and researchers I valued most highly in my Oxford lab were not necessarily the very brightest but the ones who could get experiments to work.

Cathryn then took me through some of her recent successes. Many of these have yet to be published and it would be wrong of me to disclose them here. However, I am able to mention some examples that have appeared in print. In 2014 one of Cathryn's team was contacted by a neurologist who had just seen a Hungarian Vizsla, a medium-sized hunting dog, in his clinic, with an unusual ataxia, the general name for a disorder affecting movement. The Vizsla is a rare breed in Britain so there was little chance of rounding up enough of them for a gene association study. Cathryn decided instead to go for a complete genome sequence of the one dog. She reckoned that the affected dog would be homozygous for the mutant gene and that this gene would not be found in any other breed. The trouble was that there could be very many other variants in the Vizsla apart from the mutant gene, and so it transpired. The DNA sequence picked out over three hundred variants, any one of which could be the culprit. Cathryn and her team patiently went through all three hundred variants until they found a gene that, from the deduced amino-acid sequence of the protein it encoded, fitted the bill for involvement in ataxia. Further experiments confirmed this.

The early success with the Vizsla persuaded the lab that whole genome sequencing was the way to go, and the AHT launched their 'Give a Dog a Genome' initiative. Breeders and enthusiasts from around the country were asked to share the cost of a genome sequence for a dog in their favourite breed. The initiative was very popular, and continues to be a great success, steadily increasing the number of gene sequences available for comparison in any of the lab's genome projects.

The AHT team next found the mutation for an eye disorder in a Giant Schnauzer. They followed the same strategy as for the Vizsla but with the added advantage of being able to sequence both carrier parents. This narrowed the search to variants that were homozygous in the affected dog and heterozygous in both parents. This filtered out many irrelevant sequences and made the final search considerably easier.

In another example of the success of the genome sequencing process, the AHT team found a cysteine–tyrosine substitution mutation in a form of recessive spinocerebellar ataxia in the Parson Russell Terrier.[2] This was the very same gene that was reported in four cases of cerebellar ataxia in humans. In this instance the researchers were guided to this gene by Cathryn's dog work, a rare example of success in line with the aspirations of the Dog Genome project. The mutation in humans is not the same as in dogs, which is no surprise, and will have arisen completely independently.

There is, however, one case where the mutation in dogs and humans is not only within the same gene but is the very same mutation. There is a common form of a recessively inherited eye disease called progressive rod-cone retinal degeneration (PRCD) in several dog breeds including Labrador Retriever with what I hope by now is a familiar inheritance pattern. The mutation was eventually identified, after a long hunt, in a newly discovered gene of unknown function which was named PRCD after the disease it caused. Just like the Parson Russell mutation, it is a single base change that substitutes a tyrosine for a cysteine in the protein product. As we found out in the last chapter, replacing a cysteine can interfere with the three-dimensional structure of the encoded protein and eliminates its function.

The human mutation was initially discovered in a Bangladeshi woman and has since been found in many other people. In dogs exactly the same mutation is found in all of the following breeds: Australian Cattle and Stumpy-Tailed Cattle Dog, American and English Cocker Spaniels, American Eskimo, Chesapeake Bay Retriever, Chinese Crested, Entebucher Mountain Dog, Finnish and Swedish Lapphund, Hungarian Kuvasz, Lapponian Herder, Labrador Retriever, Miniature Poodle, Nova Scotia Duck Tolling Retriever, Portuguese Water Dog, Silky Terrier and Toy Poodle.

I've given you the list in full to illustrate how many breeds with apparently little connection to each other in function or appearance must nonetheless be related back to a single dog through an impossibly complicated network of ancestors.

However, if the gene mutation is fixed in the breed and all dogs are homozygous, as in the hyperuraemia mutation in Dalmatians, then the advantage of being able to study carriers is lost. This situation could well become an insurmountable problem in behavioural traits where selection may well have driven the gene responsible to fixation. A breeder in the USA came up with an alternative and rather old-fashioned way of ridding the Dalmation breed of disease altogether. He took a Dalmatian and bred it with a Pointer. The offspring were all hyperuraemia carriers, but by crossing them with their littermates over successive generations he was able to produce a Dalmatian that looked identical to the original but did not carry the hyperuraemia mutation. Even experts could not tell them apart. Unfortunately, though, the experiment encountered stiff resistance, not because the dogs did not look like Dalmatians, but because the community of breed enthusiasts rejected them as 'impure'. Happily this resistance has diminished over the years and these dogs, free of the threat of painful bladder stones, can now be registered as Dalmatians.

Such resistance shows only too well the fickleness of some dog owners who would let dogs suffer rather than be pleased that they can own a Dalmatian with all the characteristics of the breed save one.

Although the AHT would in a perfect world like to see the elimination of all carriers from a breed, Cathryn concedes that it will be a slow process. Some breeds have very high carrier rates, for example the Shar Pei, a breed that was reduced to single figures by the end of the Second World War: 40 per cent of Shar Peis are carriers for a type of glaucoma, and even though the mutation is known, it would be a mistake not to breed from any of them. Doing so too quickly would also lose valuable genetic diversity from the breed and only encourage other recessive diseases to develop. Cathryn recommends breeding from carriers with clear dogs for a couple of generations, then gradually cutting down on carriers. With a bit of luck, half of the dogs in these litters will be clear and some at least will have all the desirable characteristics of the breed. The main thing to avoid is producing dogs with the painful glaucoma by mating two carriers together.

The wisdom of this softly, softly approach becomes clearer as tests become available for more and more genetic diseases. In some breeds there can be up to half a dozen serious genetic diseases in circulation, making it difficult to find a dog that is clear of all of them. The advice is to test all the dogs you want to breed from and avoid crossing two carriers.

Inbred populations of all animals and plants are vulnerable to effects on health. In the wild, populations of several well-known animals have at one time or another been reduced to only a few individuals. If and when the population expands from this low number the familiar problems that we have encountered with pedigree dogs begin to show. Even without specific recessive diseases showing themselves, there is a generalised phenomenon called inbreeding depression which impacts biological fitness by reducing fertility, resistance to infection and overall survival rates. For example, the cheetah population crashed about 1,000 years ago. Cheetahs, being descended from the few survivors, are all related. They suffer from high infant and juvenile mortality, low fertility and poor breeding success.

For so long as pedigree dogs are bred, the genetic issues raised by inbreeding will persist. They can be controlled but they can never be entirely eliminated. If the will is there, they can be managed. Over the last years, zoos have risen to the challenge posed by inbreeding and developed breeding schemes, including regular exchanges, to minimise the inherent risks. This becomes absolutely essential when trying to rescue species from the verge of extinction. Below a certain level of diversity, inbreeding depression makes survival almost impossible.

In the past, the criteria for selection were principally based on a dog's performance, with appearance being only a secondary consideration. Over the past 150 years this has changed as dogs have been bred to be as close as possible to the defined 'Breed Standard' in which appearance is paramount. The rewards of winning an important contest are considerable for the owner, with the possibility of lucrative stud fees to look forward to. These potential gains introduce a conflict of interest for owners. If, for example, a dog with all the right qualities is shown by a DNA test to be a carrier for a serious genetic disorder, should it be withdrawn from the competition? Owners might then be naturally wary of having their dogs tested at all, just in case. If the winner of a contest is indeed a carrier, whether or not the owner knows it, and it goes on to be a popular stud dog, then there will be large numbers of offspring, at least half of which will themselves be carriers, or worse. This happened to an Irish Setter that recently won Crufts 'Best in Show' and went on to have at least 1,000 offspring. It turned out to be a carrier for an inherited eye disease.

Homozygotes don't begin to go blind until they are about ten years old, by which time owners, and vets, are inclined to put this down to old age. I was surprised to hear from Cathryn that dogs can tolerate blindness much better than humans. She once owned a blind Retriever who could still find and retrieve a thrown ball purely by sound and smell. Nonetheless the issue of the popular sire who carries a genetic disease is a serious one.

Ultimately the success or failure in improving the genetic health of dogs depends on the breeders. The exactly analogous human situation is the successful elimination of Tay-Sach's disease in the Ashkenazi Jewish population, achieved by leadership and determination. The Kennel Club can lead by example but it has no statutory powers. Over the past decade it has risen to the challenge of tackling the issues raised by the BBC documentary by, among other things, funding the Animal Health Trust research effort. The Kennel Club has put its money where its mouth is. It has to work to improve the welfare of dogs by persuasion, with very limited recourse to legislation. I don't envy them their task.

21 The Scientist Who Came in from the Cold

A now-famous experiment in Russia had anticipated the same conclusion as the US team about the mechanism of domestication, but without the fearsome armoury of modern genetics. The experiment was noteworthy on its own, but perhaps even more remarkable was the fact that it was carried out at all. The scientist Dmitri Belyaev began working when Stalin's regime was at its height. Stalin had appointed a Ukrainian, Trofim Lysenko, the son of a peasant, to dramatically increase Soviet agricultural production following the widespread crop failures that followed the collectivisation of farming in the early 1930s. Lysenko had an idiosyncratic approach to his task that rejected established genetic theory, believing instead that genes were unimportant in determining crop or milk yields. He insisted that improvements could be made only by altering the way in which animals or plants were treated. In the balancing act between nature and nurture, nature was ignored altogether. As Lysenko grew in prominence and influence, many of his opponents were removed from their positions, imprisoned or even executed. Genetics became a dirty word in Soviet Russia, textbooks were removed from university libraries and geneticists themselves became enemies of the State.

It was under these unpromising, to say the least, political conditions that Belyaev began. He worked in Moscow on a dissertation incautiously entitled 'The variation and inheritance of silver-coloured fur in Silver-

Black foxes'. Belyaev survived Stalin's regime and in 1959 moved east to the Siberian town of Novosibirsk where he remained for the rest of his working life.

Soon after he arrived at his new laboratory he began his most famous breeding experiment, which he was able to disguise as a purely commercial exercise and thus avoid the long reach of Lysenko and his cronies. Silver foxes had been reared on Russian fur farms since the nineteenth century and fetched a good price. They are called silver foxes because of the white guard hairs which protrude beyond their otherwise blue-black coats, but they are in fact a colour variant of the familiar and widespread red fox *Vulpes vulpes*. Belyaev disguised his research as being designed to increase the proportion of silver guard hairs and thereby put up the value of the pelt. He began his selective breeding programme with that declared aim, but it was, in reality, purely Darwinian in design.

Instead of choosing to breed from foxes on the basis of their coat-colour, he chose a completely different criterion. He would select not on the basis of a physical characteristic at all, but instead on a single variation in behaviour, namely tameness. He began with thirty males and a hundred females from a commercial colony in Estonia where the animals had been bred for fifty years without selection of any kind. Belyaev was focusing on the foxes' reaction to humans. They were still wild animals at heart, even though their ancestors had been in captivity for so long. Most of them reacted to humans with a mixture of fear and aggression, retreating to the back of their cages whenever approached and biting when anyone attempted to handle them. As you would expect, there was a variation between individuals from the colony, and about 10 per cent of them were noticeably more tolerant. The stage was set and the experiment began.

When the cubs were about four months old, Belyaev or one of his assistants would try to stroke them. This was repeated for the next three months. Belyaev chose the least aggressive animals to breed from, and these formed his experimental group. He was careful also to have a control group in which mating was random. After only twenty generations of selection the experimental foxes were showing consistent changes in their behaviour. They became affectionate towards humans almost as soon as they opened their eyes. They became playful and enjoyed the interaction with their keepers in ways that would have been unimaginable at the start of the experiment only twenty generations before. The extent of the behavioural transformation was remarkable, but less so than the speed at which it had occurred. Evolution under natural selection is an excruciatingly slow process, taking thousands or even millions of years to produce substantial changes of the magnitude shown by the foxes in only a fraction of the time. There were other surprises.

Dmitri Belyaev with his selectively bred foxes. Photographed in Novosibirsk, Russia, in 1984.

Even though the criterion for selection had been entirely behavioural, after a while the experimental group of foxes began to exhibit physical changes. Their ears became floppy, their tails curled and their coats became uneven, with patches of different colours. The skull shape changed too, becoming broader and with a shorter snout, both features usually confined to immature animals and oddly reminiscent of the differences between wolf and dog.

Brian Hare, the expert on animal cognition, wanted to see whether these collateral changes associated with domestication extended to the ability to read human gestures. One summer he and his research associate visited the Novosibirsk facility and put the experimental foxes through their paces. He used the familiar baited cup trial and was astounded to find that the experimental foxes were absolutely brilliant at it, scoring far higher than by chance alone. It was hard to design the experiment to test the control group because the foxes were so skittish. Through a series of ruses, they were persuaded to perform, which they did rather badly. Along with the behavioural changes easily attributable to the single selection criterion, tameness, it seemed as if the assimilation process brought with it a genetic package that included some physical changes and cognitive abilities that we can easily distinguish in the domesticated dog.

The inference to be drawn from Belyaev's work is that some of the physical changes associated with domestication do not have a direct genetic basis but are the effects of secondary hormonal adaptations following on from the selection for tameness.

Even though Belyaev's research was carried out on foxes, the results may help explain the biological mechanism behind the human–dog bond. In 2015 a Japanese research group found evidence for the involvement of the hormone oxytocin in promoting the bond between the two species based on mutually understandable communication. Oxytocin has been called 'the love hormone' because its action serves to promote feelings of attraction and attachment between individuals. This appears to be a secondary effect of the hormone, whose principal activity is to dilate the cervix during childbirth.

Oxytocin is a short neuropeptide of only nine minor acids and is made in the hypothalamus deep in the brain behind the eyes. In a display of thrifty evolution, oxytocin orchestrates an impressive repertoire of different activities surrounding sexual reproduction, the close attachment we call 'love' being one of them.

All mammals use oxytocin for much the same reasons, but because human babies develop so slowly compared to other animals and are dependent on their parents, and particularly their mothers, for many years, the 'bonding' effects of oxytocin are especially strong.

It was always assumed the feelings of attraction and attachment induced by oxytocin only applied to members of the same species. However, in 2015 the Japanese scientists published a startling paper contradicting this long-held assumption by showing that oxytocin is closely involved in the strong bond that humans have developed towards their dogs – and vice versa.[1]

The scientists measured the oxytocin levels in thirty owners and their dogs, before and after they engaged in what was described as 'mutual gazing'. It has been known for a long time that looking closely into each other's eyes is a key factor in establishing the strong attachment of a mother and her baby. The Japanese study was the first to show that the same hormonal mechanism, involving oxytocin, is also behind the attachment of a dog and its owner. Levels of oxytocin went up after a session of mutual gazing, not just in the owners but in the dogs as well.

The same responses did not occur in encounters between humans and captive wolves, even those that are accustomed to close human contact. It seems likely therefore that the strength of the human–dog bond, mediated through oxytocin, is a direct consequence of the selection process that turned wolves into dogs.

Oxytocin itself has been used in the treatment of autism and of post-traumatic stress disorder, and it may well be the case, though this is yet to be proven, that therein lie the benefits of assistance dogs for these and other conditions. The finding that dogs also experience an oxytocin surge after mutual gazing suggests that the 'love' is reciprocated rather than being a ploy on their part to bring out favourable behaviour in ourselves.

Belyaev died in 1985, but his long-time associate Lyudmila Trut has carried on the experiment, which is partly financed by the sale of her remarkably affectionate foxes as pets.

The Autumn Muster

I have lived with DNA and evolution for most of my professional life as a scientist and so felt adequately knowledgeable to write this book. But I soon realised, as I began to write, that there was a risk that the book would be missing the central character – the dog itself. Fortunately help was at hand. My spirited wife Ulla grew up in the Danish countryside surrounded by dogs and carries none of the prejudices instilled in me ever since my childhood experience with the Hell Hound down the road. Ulla loves dogs and she shows it. To augment the scientific element that makes up the bulk of this book, Ulla interviewed a selection of owners and their dogs, and we will come to the interviews in the next chapter.

Meanwhile, during a spell in New Zealand, I had arranged to visit a man who worked with dogs for a living. I imagined he viewed his dogs not as pets but as working companions, very much as they had been for thousands of years.

Ewan lives and works in Mackenzie country, the high plateau on South Island in the shadow of the Southern Alps. There are few sights in the world to compare with the snowcapped peak of Aorangi, the highest in the country, viewed across the turquoise blue of Lake Pukaki. Some mountains are terrifying to behold, some peaks are disappointing when you see them, some summits are hidden behind the folds of surrounding hills, but Aorangi, the Cloud Piercer, is simply beautiful. It stands alone, a glistening meringue of snow and ice floating between azure lake and cobalt sky.

This is where Ewan works his dogs. Like most New Zealanders he is not at all sentimental. He knows he lives and works in stunning surroundings but, after ten years of mustering (rounding up sheep) in the high country, he takes it all in his stride.

The New Zealand economy relies heavily on selling its agricultural products all around the world and, since the first cargo of frozen lamb reached England as far back as 1882, there has been a flourishing export trade in meat as well as in wool. Official government statistics show that sheep still outnumber resident humans in New Zealand, but the proportion has been declining since it peaked in 1982. Then, there were 70 million sheep and 3.18 million people – that's twenty-two sheep per person. Since then, a combination of factors – including lower international wool prices and a switch to dairy cattle to satisfy the Chinese demand for powdered milk – has reduced the sheep population to just under 30 million by 2015. At the same time, the human population has grown to 4.6 million, lowering the sheep/human ratio to just over six sheep per person.

To gather in these vast flocks, there are about 200,000 working dogs in New Zealand. Without them, it would be quite impossible to muster the extensive areas of high country. Needless to say, these working dogs are a world away from the pampered pooches that some of their cousins have become. No professional dog walkers for them when their owner is too busy to take them out for their exercise in the park. I was keen to find out about the relationship of these working dogs with their owners, the sheep station managers and the professional musterers.

I had arranged to meet Ewan and his dogs at his home near Omarama. A few days later, he phoned to say he had a contract in the nearby Upper Ahuriri valley and offered to take me along. I jumped at the chance. This was the autumn muster, when ewes and their lambs are brought down from the hills. Once down, some lambs are kept for fattening on the rich valley paddocks, others are sent directly for slaughter, and the ewes are made ready to be tupped to begin the lambing cycle all over again.

The muster began early, just after dawn, on a chilly February morning. Like many valleys that radiate from the backbone of the Southern Alps, the Upper Ahuriri is steep-sided but with a wide flat bottom, the result of millennia of erosion and accumulation of thick layers of glacial gravels. It is flat enough for an unsealed road to reach 35 kilometres into the mountains until it reaches the wall of peaks, still covered in glaciers, that separate the Ahuriri from the Hunter valley to the south. There are three sheep stations in the Ahuriri valley. Two are active, the third was purchased by the Department of Conservation in 2004, cleared of livestock and is now a sanctuary for one of New Zealand's rarest birds, the *kaki* or black stilt. Today we are mustering a run at the lower end of the valley belonging to the Ben Avon station.

As dawn was breaking, we met up at the old wooden bridge over the river. The bridge was just strong enough to take Ewan's pick-up. While I waited for him I glanced over the white-painted rails into the clear blue water below, its unreal colour the result of light scattering by suspended particles of fine glacial flour ground from the mountains. There beneath the bridge, close to the bank decorated by dancing patches of viper's bugloss, the same big trout that I had tried and failed to catch a dozen times on earlier visits lay suspended in the current. This time I wasn't even going to try to catch him.

Soon a cloud of dust billowing up from the main track announced the arrival of Ewan and his dogs. A few minutes later we were joined by two other musterers and their dogs. I unlocked the gate at the far side of the bridge, then jumped into the passenger seat of Ewan's truck and we were away. Up and up the snaking zigzags of the rough drive track we went as we pushed towards the ridge. A rare New Zealand falcon, beautifully marked in streaked dark brown and cream plumage, dashed across our path as we passed beneath a rocky outcrop. We drove higher and higher past erect spikes of grey and yellow woolly mullein lining the track, and as we did so new peaks appeared on the horizon. At the head of the valley was Mount Barth, its permanent snows dazzlingly white in the strong sunlight. As we reached the ridge 600 metres above the valley floor, the giants of the Southern Alps lined the far horizon. Mount Sefton, Mount Tasman and in between them Aorangi, its perfect summit reaching beyond 3,600 metres above sea level. What a place to work!

Ewan took out a pair of binoculars and scoured the slope above us, looking for the sheep. These were merinos, the hardiest of all New Zealand breeds. They needed to be tough to survive the wind, rain and snow that batter the high country for a large part of the year. The warmth of merino wool is legendary, but the sheep also make a lot of lanolin to waterproof their coats against the elements. This means they easily pick up dust, which makes their coats look a dirty grey. It also makes them very hard to see – hence the binoculars. I had a lot of trouble picking them out, even with binoculars, but Ewan quickly assessed their distribution on the slopes.

The station manager had told Ewan that there were 1,200 ewes with their lambs. With a lambing ratio of 1.25 this year, that meant the muster was bringing in almost 3,000 sheep. They were spread around a vast bowl of mountainside rising, maybe a kilometre or so away, to a skyline of jagged tors like the backbone of a half-buried stegosaurus. These rock formations are typical of the schist high country where rain, wind and frost have worn down the softer rocks to leave the vertebral plates exposed.

The ground alongside the track was now a jumble of matagauri shrub bristling with lethal spikes, evolved as protection against browsing moa. The moa is long gone, but the spikes remain. Above the matagauri, clumps of snow tussock gave way to steep and broken scree below the tops. Ewan and his friends plotted the muster. Brian and his dogs would take the southern side of the run, Iain would take the centre, leaving

Ewan to cover the northern flank. We travelled half a kilometre further up the track and Ewan released the dogs from their cage in the back of the pick-up. They were excited and, so it seemed, looking forward to getting the sheep rounded up just as much as Ewan. Among the pack, for that is what it was, were two black-and-tan Huntaways, a powerful thick-set New Zealand breed raised for just this type of country. Though no one was completely sure of their origin, there was reputed to be a mixture of Labrador, Rottweiler and Dobermann Pinscher somewhere in the ancestry. Owners of working dogs take pride in never entering them for shows and do not concern themselves with pedigrees. All that matters to a musterer is that his dogs are keen workers, which Ewan's clearly are.

Huntaways control sheep from a distance with their deep bark and intimidating presence. With a single piercing whistle from Ewan the two Huntaways ran fast left and right up towards the skyline, covering the ground in a series of long bounds. Their job was to prevent the sheep from scattering and disappearing over the ridge. Once positioned above the flock, another whistle from Ewan and the dogs began to bark, deep and loud. The sheep looked up as the dogs trotted along the ridge over the rocky tors, barking as they went and forcing the sheep down the slope. Then Ewan released the four Border Collies and, on his whistled command, they dashed uphill to surround the flock and cut off the retreat. Keeping low and worrying the outliers with short barks, the Collies slowly concentrated the flock towards the central gully of the mountain bowl. As they came lower and lower, the density forced them into a tighter and tighter group that, it seemed to me, began to move as one, swirling like a shoal of herring harried by dolphins. The two Huntaways rounded up the few stragglers into the main flock which was by then moving slowly towards Ewan's truck, kept in tight formation by his energetic Collies.

A group on the right tried to break away and, immediately, Ewan whistled to the dogs to cut them off. Ewan and I got back in the truck and retreated part way down the track. The sheep, hemmed in by the dogs, followed like a great grey river down towards the valley floor. Wheeling overhead, a kea, the indigenous New Zealand mountain parrot, let out a raucous shriek as if to say 'good riddance' to our unwelcome intrusion into the peace and quiet of the high country.

The pens were 3 kilometres upstream of the old bridge so we prepared to ford the river. The Ahuriri, like so many South Island rivers, is braided. They do not follow discreet and permanent courses, but are constantly changing their route through the gravel beds in the valley bottom. Floods from the spring snowmelt shift the stones and force the river to move sideways, creating new channels and closing off old ones. Only where there is solid bedrock and the river is forced through into a narrow channel can a regular bridge be constructed. Elsewhere, bridges are sometimes a kilometre or so long, straddling the gravel channelled by the river, though there are none across the remote Ahuriri. In time, these bridges will all have to be moved when the rivers change course beyond their present limits.

Had there been a lot of rain in the hills, fording the river would have been too dangerous and the muster would have had to be postponed. Fortunately for us, it had been dry for a week and the river was running low enough for us to use the traditional crossing point. Sheep are surefooted but timid, and the river crossing could not be hurried. While the Huntaways guarded the rear of the flock, cutting off their retreat, the Collies patrolled the margins. The sheep hesitated at the water's edge, nervous of taking the plunge into the icy current. It was vital not to panic the flock, and the dogs remained quiet.

After what seemed like several minutes, pressure from the back of the flock nudged the first sheep into the water, and once that had happened the whole flock trooped across the river. It was shallow here and they did not have to swim. Two of the Collies ran across a little way upstream and made sure the emerging animals kept to the track. Such was the size of the flock, it was a full twenty minutes before the last sheep was across. A little later they reached the paddock and Ewan closed the gate behind them. They soon settled and were eating the fresh green grass. A little later first Brian's then John's flocks were in the paddock. The whole muster had taken nine hours. Even though I had done very little other than watch Ewan and the dogs, I was exhausted.

Once we had secured the paddock gate and handed over to Bruce, the station manager, we loaded up the dogs and set off on the drive home. Ewan's house was a comfortable bungalow surrounded by 20 acres of paddock with a few cattle, a chicken shed and a couple of dozen sheep which he used to train his dogs. Ewan was a professional and took on all sorts of work. One day he might be helping with a high country sheep muster, as today. On another he might be rounding up cattle or deer on the flats. When he was not out working the dogs on the land, he was judging sheepdog trials. The active outdoor life clearly suited him. At 70, he was as fit and lean as a 30-year-old.

In a shed round the back of the house Ewan showed me two young Border Collie pups, two months old, part of litter of six that he had decided to keep and train himself, selling the others on. In another pen were two slightly older Huntaway pups that he had also raised. He was going to keep one and had already sold the other to a farmer in Wales. When it was a little older the pup would be flown halfway round the world to his new home. Ewan's dogs were highly valued and valuable, going for anything between NZ$5,000 and NZ$10,000 each.

Ewan had twelve other mature working dogs, Collies and Huntaways, and they were kept outside, each in its own plastic barrel, laid flat and with the lid cut off. At first sight, these might have looked unnecessarily spartan and they certainly contrasted with the upholstered dog beds I had seen at Crufts. However, they were entirely waterproof and, with a lining of straw, snug and warm.

Ewan's dogs were his livelihood and not his pets. In order for them to work as a team he had to maintain a position of dominance. If he did not, the dogs would be useless on the muster and would run off after the first rabbit they saw, never to be seen again. It was essential that they obeyed his instructions. He was their leader and they were his pack.

What interested me most about Ewan and his dogs was how he felt about them. Did he love them? Before I met Ewan I had supposed the answer would be 'No' and that he would just view them as tools of his trade. But when I asked him this question directly he said 'Yes' immediately. I asked him if he could say why, a question that flummoxes most pet dog owners who, though quick to declare their love for their dog, cannot put their reasons into words.

Ewan's feelings for his dogs had nothing to do with transference and he had no difficulty reeling off a string of reasons for his affection. They had given him great pleasure, to say nothing of providing a comfortable livelihood for him and his family. They had given him a life in the open air, his days in the high country with just him and his dogs among some of the most beautiful scenery on earth. He was very proud of them and, he told me, his greatest satisfaction was to bring a dog on from a puppy, watch it grow and improve through training and finally see it take its place with the others on the muster. Not all dogs made the grade and he sold them on, often as pets. That's also what happened when a dog grew old and was no longer physically capable of the gruelling workload. Ewan always found them homes in which to spend their retirement. He has never had a dog put down just because it was unfit through age or illness. He respected his dogs and, yes, he certainly loved them. Whether this was reciprocated is of course impossible to know for sure, and is in many ways irrelevant. They made a good team.

I said goodbye to Ewan and set off for home over the Lindis Pass. It was early evening as the car climbed to the summit, and the low sun spread a gentle drape of bronze velvet over the hillsides, picking out the grassy tussocks like giselé tufts on a Tudor doublet.

23

The Girl Who Talks with Dogs

Ulla's interviews with dog owners follow what I call the 'Easy Rider' approach. There is no plan except just to get started and see where things lead. Ulla set off with a voice recorder and did exactly that. The recordings were so full of laughter they took an age to transcribe, so please imagine the hilarity that peppered the words but could not easily be transferred to the page. It certainly did help the dogs and their owners, most of them strangers, to relax. I certainly could not have done it.

These interviews do not by any means constitute a statistically significant sample, being recruited largely by chance encounters in Hyde Park close to Ulla's London studio. I should have said that Ulla is a painter and needs to be close to the galleries that display, and occasionally sell, her work.

The only dog with which I can claim any sort of friendship is Sergio, a boisterous Italian Spinone who lives on the Isle of Skye. Ulla and I stay on Skye for much of the year, and Sergio belongs to a friend and neighbour a couple of houses along the road. Sergio is nine years old and, although not a huge dog, is certainly large and powerful, so when we were first introduced I was always careful to stand in the background. Ulla on the other hand had no hesitation in running her fingers through Sergio's thick coat, not always managing to avoid the threads of drool that dripped continuously from his mouth, a characteristic of this breed. Now when

Ulla Plougmand, the interviewer, with Sergio (Italian Spinone).

we come back to the island Sergio seems to know we have arrived and runs up to our gate and launches into an extraordinary performance, running around in small circles while softly singing with muffled bleats. Though with time Sergio has got used to me, there is no mistaking his disappointment if I am there alone; he quickly loses interest and trots back down the road.

Bonnie and Bernie

Not long before I was due to meet up with Ewan, the professional Kiwi sheep musterer, to ask him about life with his working dogs, Ulla had a chance to interview Bonnie and her new owner Bernie. They lived in a spacious lodge a short way outside Wanaka on the South Island where we had been staying. Bonnie lived up to her name. She was a very pretty 4-year-old Collie crossed with a Kelpie, an Australian sheepdog, with long legs which gave her an elegant, bouncing gait not unlike the wolves we had encountered when we met up with Shaun Ellis the previous year. Bernie and her husband Paul had lost their mischievous Springer Spaniel, Minnie, two years before and felt it was time to replace her. She had been such a character with a most charming way of begging for food – she would pad up to the table and sit quite still just looking at the guests with doleful eyes whose message was clear and could not be ignored. Minnie's patience was always rewarded by a piece of cake or some other tasty morsel. But now she was gone. Bernie and Paul did not want to replace her like-for-like, so they looked around for a completely different dog. They found one on Forever Homes, a New Zealand-based website devoted to rehousing all sorts of unwanted animals. They liked the look of Bonnie and arranged to visit the lady who was caring for her after she had rescued the dog from a farm.

'We had been looking for a dog for some time,' explained Bernie. 'We always knew that when the right dog came along we would know it, so we arranged to go and see Bonnie to see if she was that special dog.'

'And was it love at first sight?' enquired Ulla.

'Well. Yes. But we knew she would be a challenge.'

'Yes, the thing you notice straightaway about Bonnie is that she is obviously extremely shy. What did you learn from the lady who had rescued her that might explain this?'

'The lady who was looking after her had a farm and she had rescued Bonnie from another farm nearby. She had been brought in as a working dog on the sheep station, but she just didn't want to work. Farmers in New Zealand, like everywhere else, are not sentimental. She was costing money to feed so they tried everything to make her work. They beat her and when she still wouldn't work she was left chained up.'

'It is so heart-breaking to hear a story like that.'

'She doesn't bark because it was beaten out of her. She was meant to be a heading dog – they are quiet and they creep along close to the ground. Farmers routinely beat the bark out of them from when they are puppies, or they put "bark-collars" on them so that every time they bark they get an electric shock or the collar emits a high-pitched whine which drives the dog mad.'

'But now she has a happy life, doesn't she?' declared Ulla. 'I took a walk earlier today and she has a lovely little house surrounded by grass and flowing fields to run around, everything a dog could want.'

'Yes, but I didn't realise until I spoke to a farmer that I should not expect her to do a lot. Farm dogs go out and work hard for maybe half an hour then they'll sleep. Then they might do a bit more work, but they are mostly just sitting around waiting.'

'So they have sudden bursts of energy and then they are knackered out and they need their rest. Good God, it sounds like me you are describing. But now she is settling down, is she becoming less of a farm dog and more part of the family?'

'Yes, she comes into the house and goes anywhere she likes. She has her crate in the corner over there which she goes into at night. She used never to come into the sitting room when the television was on, because of course she had never seen one before. When she first came to us she was scared of just about everything. She used to run off when the fridge door was opened. I would show her and say

"Look, it's the fridge door making that noise." And the next time she heard it she was fine.'

'Because she is bright and she learns very fast,' said Ulla.

'At night when she comes in she straightaway jumps onto our bed.'

'But you and Paul are very hands-on people, which I love, and lots of people would say "Oh No. Not on the bed." I used to sleep with my Cocker Spaniel *in* the bed. In fact, I was pushed over into a corner.'

'She lies on the bed between the two of us, on her back with her paws up in the air. She loves to have her belly rubbed. Then when it's time to go to sleep we turn off the light and say, "Go on then, off to bed", and she jumps up and goes straight to her crate.'

'I think that's really lovely and I'm surprised to hear that you have got that far with her in, what is it, only three months.'

'Yes, but she is a very bright dog, which I like, but even with the bad treatment she received on the farm you can still see the sheepdog instincts coming through. For instance, if she is on the other side of the paddock and I call her she will run straight back and sit down beside me. Minnie never did that.'

'So, Bonnie is still a sheepdog at heart, but a much happier one than before,' Ulla concluded.

Sheba and Alexander

Alexander Thynn, Lord Bath, is an interesting, unusual and some might say eccentric member of the aristocracy. I met him first in 1997 when I was attempting to recover DNA from human fossil bone. His large estate surrounding Longleat, the beautiful Elizabethan house in Wiltshire in the south-west of England, extends to the famous show caves in Cheddar Gorge. In one of these, Gough's Cavern, two human skeletons were recovered in the last century and I managed to recover some DNA from them. I needed Lord Bath's permission to do this, as he was the owner of the cave, and he became very interested in what I was doing, especially when I found a genetic match between one of the skeletons, the so-called Cheddar man, and his butler Cuthbert. When I went to see him at Longleat to go over the results, he was accompanied everywhere by his cream Labrador called Boudicca. Perhaps unconsciously already planning this book, I took a DNA sample from her. Nothing came of it because I used a DNA technique closely modelled on humans, but it was enough that Lord Bath remembered this and years later graciously allowed Ulla to interview him. By then, Boudicca had died and her place at Longleat had been taken by Sheba, another Labrador with lots of Poodle mixed in. The interview took place in Lord Bath's apartments on the second floor, a space once taken up by his father's collection of Nazi memorabilia. When Alexander succeeded to the title on his father's death, the collection was the first thing to go and Alexander converted the space into comfortable living quarters with a grand view of the lake and wooded parkland stretching into the distance. As befits the owner of a large country estate, there have always been dogs in the family, but as you see from this account that does not diminish by any means the love between a dog and its owner.

Ulla began her interview with a very direct question.

'Alexander, can you tell me why you love your dogs. What I mean by that is what characteristics do you appreciate most in your dogs?'

After a thoughtful pause he replied:

'With Boudicca it was loyalty and love. With Sheba I like being taken as part of her background which she knows is safe. Now that Boudie is no longer here I think Sheba realises there's a gap in the requirements which she is expected to fill.'

'I noticed something over lunch. Boudicca because she was a Labrador she could not stop eating and you would become her best friend if you had something to feed her with. I noticed that Sheba is much more withdrawn that way.'

'Yes, she is not going to be a fat dog, but she is not going be a starving dog either.'

'She is a lucky dog actually. Would you say with your dogs that you treat them as part of the family?'

'Well, we are a sort of bunch all together.'

'My old Cocker Spaniel that I had as a child used to sleep in my bed. I did not mind a bit. Would you dream of having your dog sleep on, or even in, your bed?' asked Ulla.

'They wouldn't have it any other way. She comes up rather aggressively and lies on top of me for part of the evening and then when I start wriggling a bit she gets impatient and moves a couple of paces away.'

'So in a way she's taking it for granted that it is her bed as well. If anybody had to move, it might be you!'

'The expectation might be there but she would be surprised at the resistance. At mealtimes Sheba will come and put her head on my knee to remind me that she is still there waiting, quite patiently, for her due desserts.'

'She's very well-mannered, your Sheba. Now, with all this enormous love between a dog and its owner I wanted to ask you what you would do if Sheba became unwell. I knew a dog called Timmy, a Labrador like Boudicca, that I used to take for walks in the park. When he was about twelve years old, his hip had gone and the lady who owned him had a hip replacement done for him. What lengths would you go to if your dog needed something like that?'

'If my dog is suffering pain I'd take that as a decision made for me. She mustn't be in pain and we have to provide a way of her getting out of it. I had already taken that decision with Boudicca. She could no longer get up, and though she was not in pain there was really nothing left. When she got very old and could not move any more, her quality of life was such that I would not prolong it.'

'You obviously loved Boudicca. It's now possible when a dog dies to clone it and, in a way, get an identical replacement. Would you ever consider doing that?'

'I've been lucky with whichever dogs I have chosen and so the need for cloning really hasn't been there. I look forward to the next dog with its own personality and to surprise me in some way. I'm not really for having an identical dog, although with Boudicca it would have been lovely to have her back again. You take a gamble with any new dog on how they will turn out. I suppose it's the same with children.'

'Has there ever been a gap between the dogs, or has there been a time where you've said that if ever a dog has passed away or you have had to let one go, that you won't get another dog?'

'I've had dogs all my life, as a child and onward, so when I lose one there may be a short gap but I soon begin to think about what kind of dog I'll get next. When Boudicca died I decided to try a cross between a Labrador and a Poodle, a Labradoodle, and that is Sheba. With Sheba, she is more Poodle than Labrador so if I have another dog in the future I might go back to having a Labrador again, even if it loves its food more than it loves me.'

Elton and Ulf

Ulla was visiting Palma de Mallorca when she came across a dog boutique in the centre of town near the cathedral. Attracted by the window display, she walked in and was soon talking to Ulf, the Swedish owner. He specialises in high-quality photographic portraits of dogs, and Ulla began by asking him about the sort of people who commissioned them.

'We have been quite overwhelmed by how much people are willing to spend on their dogs these days. How much do you charge for a portrait of a dog?'

'So, if you want me to make a portrait, a small one would be around €150, going up to about €600 for the biggest that I do. This one here is Elton, my own dog. Elton is a Jack Russell terrier. He was the inspiration for me to open the shop. He is quite old now, thirteen years, but with luck he will live to be at least sixteen if he stays healthy.'

'You mean Elton, like Elton John?'

'Of course, I was going to say that but I didn't want to. There is only one Elton.'

'Will you show me some of the other goodies you have in the shop? For instance, what are these?'

'These are little models of dogs hand-carved in special wood. This one is made from walnut. Here is a Yorkshire terrier and a King Charles spaniel and they are €499. For a bigger dog obviously the price goes up, to over €1,000 for a really big dog. But most people want the models of smaller dogs so I don't have any big ones in the shop.'

'They are really beautiful quality. Where are they made?'

'These come from the UK, like a lot of the items in the shop. Germany is also another big producer of dog accessories like these beds. These are made completely from organic fabrics and stuffed with natural latex. These, over here, are made from Harris Tweed, which means they are coloured by hand. They cost €200 for a small dog up to €500 for a big one like this.'

'It's huge. It's like a swimming pool.'

'Yes. You could have it on the floor in front of your sofa. They are very comfortable. I sold one to somebody who did not have a dog and she tells me she often sleeps in it herself!'

'And look at these lovely collars and leads. You could rig up a boat with this one and sail the Atlantic Ocean!'

'Yes. It is made from a real yacht rope with stainless-steel fittings. It is handmade in the UK and costs about €100 for the rope and a matching collar. So quite cheap really. You can jump up a step with this lead, which is actually made from German boat rope from Hamburg and you can choose the colour and the thickness. A rope and collar would be €150. It is guaranteed never to break. It doesn't get old. It just matures.'

'Here on the table we have keyrings and mugs with doggy pictures and here are some birthday cards. I will buy some of these because I am the Danish aunt to several dogs. And here you have bottles of wine with dog pictures on the label. Surely these are not meant for dogs to drink?'

'No. These are bought as presents for dog sitters when you go away on holiday. All the labels are hand drawn by a gentleman from Palma. It is good wine and I charge €90 for a bottle.'

'And this metallic dog. What is this for?' asked Ulla.

'It is actually a coffee grinder. I just thought it was funky. They cost from €150 to €300 depending on the craftsmanship. I am also working with a girl in New York who designs clothes for dogs. I would not normally dress my dog, although Elton is getting quite old so I cover him up in the winter. But there is quite a lot of demand now for dog clothes from top designers like the lady from New York, so I think I will start to stock a few. Also I am getting some Fairtrade clothes made in Nepal. They make beautiful hand-knitted sweaters, all made from local wool. Nepalese sheep have to live in the mountains so their wool is very warm to wear. It is my shop so I can stock the items I want. I don't have to make millions to get by.'

Roo and Stephen

Ulla met Roo and owner Stephen in Hyde Park. Roo is a beautiful 6-year-old black and white spotted Dalmatian, a breed of dog with close historical links to the age of stagecoaches. The distinctive Dalmatian was a favourite dog whose job it was to run alongside the stagecoach and protect it from highwaymen.

As well as being a family pet, Roo is also a canine supermodel, having featured in advertising campaigns for, among others, Fortnum & Mason, the top-notch grocers in Piccadilly, and the fashionable designer Cath Kidson. But unlike her human counterparts, Roo has not allowed her celebrity to go to her head.

'Roo is a very friendly dog with a lovely character. She is always happy to see you and greets you with a kind of smile and a little kangaroo hop when you call her by name. And she is extremely loyal,' explained Stephen.

'So would you say that you love Roo? It is a silly question, isn't it, but Bryan is a scientist.'

'Yes, definitely, is the answer I would give. She gives me unconditional love and I try to do the same in return. In the park she will welcome and greet anybody, but as soon as I blow her whistle she knows to return. Then I give her a small treat or sometimes a carrot.'

'Do you consider Roo to be a member of the family?'

'Oh, certainly. She is very much a member of the family. She has space on the ground floor and she is not allowed to go upstairs to the bedrooms. She has her own bed under the stairs and that's where she rests and sleeps. That is her domain.'

'So Roo doesn't get to lie on your bed, for example?'

'No. Her territory is downstairs and she knows it. We have fairly strict rules about what Roo can and cannot do. We do not feed her at the table and she eats from her bowl in the utility room. Sometimes we give her leftovers but not at the table. They go into her bowl.'

'You said that you feed Roo with carrots in the park, so is she mainly vegetarian?'

'You have to be careful what you feed Dalmatians. They have problems with meat like beef because of the purines, so her diet is mainly vegetarian. We give her dry food which contains chicken and corn and maize and that sort of thing which is made specially for the breed.'

'Where did you get Roo from?'

'We knew we wanted a Dalmatian but you have to be careful. Someone recommended a breeder on the West Coast and we drove up there to look at the litter. They were all cute but my wife and I were both drawn to the same puppy. Dalmatians can have hearing problems, and problems with their eyes, so we sort of checked for that and so far she has been fine. They are not a very popular breed at the moment, unlike when the film (*101 Dalmatians*) came out in the 1960s. I don't really know why that is. Dalmatians originally came from Croatia, where they started as hunting dogs. They are very strong and very fast. She can outpace a Greyhound over a long distance.'

'There seems to be no limit to what people are prepared to spend on their dogs. Bryan and I have just been to Crufts and we saw a huge range of accessories for sale, including padded leather dog sofas for £500 or more. Would you think about something like that for Roo?'

'No. At the end of the day she is a dog so let's keep it in proportion. She is a great companion and family friend. When she isn't here or in kennels we certainly do miss her. So yes there is a massive connection. Obviously she has her check-ups at the vet and all the necessary jabs, but I draw the line at spending money on special furniture. It's big business here in the UK, one of the most successful, but I would not spend large amounts of money on Roo unnecessarily.'

'What if, God forbid, Roo became very ill and needed a hip replacement for example?'

At this point Stephen's wife enters the room. She speaks very fast in an accent which Ulla recognises as Swedish.

'It's always nice to meet a fellow Scandinavian, especially one with a lovely dog like Roo. I was just asking Stephen what you would do if she became seriously ill. Would you pay for treatment or would you have her put to sleep?'

'Roo is only six years old so there is plenty of life in her so, depending on what the problem was, I very well might pay for treatment. But if she was old and had no prospect for a life of good quality and if the vet recommended it, I think it would be best to put the dog to sleep. Of course, we would all be extremely upset because we love our dog, but on the other hand we would not want her to suffer.'

'That would be cruel,' said Ulla. 'What a pity we humans do not have that option at present in our country.'

Sergio and Innes

For the past twenty years, I have lived for part of the year on the Isle of Skye off the north-west coast of Scotland. Without a regular dose of the wild I am certain I would have cracked up long ago. My small house on the island once belonged to the great Gaelic poet Sorley Maclean and, as I write, I imagine that some of his spirit diffuses from the fabric. The house is in a small clutch of crofts, hardly a village, close to the sea and dominated by the granite bulk of Glamaig, the highest of the Red Cuillin.

My neighbours a couple of houses along are Innes and Anna, along with their four daughters whom I have watched grow from small children to beautiful young ladies. Like many islanders, Innes does a variety of work including building and repairing mountain paths. He is also a musician with a Gaelic band that plays all around the world. His dog is a 9-year-old male Italian Spinone called Sergio and is the only dog for whom I feel any affection. I am slowly getting used to him and have even begun to take him for walks through the heather on the hill behind the house. Mind you, I don't think he is all that bothered about me. Ulla, of course, fell in love with him the first time they met. The feeling was mutual so it was she, not me, who asked Innes all about Sergio.

'Innes, I have known Sergio since you got him, but let me ask you what your reasons were for getting a dog in the first place.'

'There are a couple of answers to that question. At the time Anna and I were talking about getting a dog, Anna remembered that her

Sergio, the boisterous Italian Spinone, at home on the Isle of Skye.

sister had a friend with a litter of eleven Spinone puppies. She was desperate to find homes for them all and we were asked if we would take one. It seemed impetuous at the time but we thought, well, we could at least go and see the dogs, which we did, and said yes.'

'It's a big responsibility.'

'It is. But we soon realised that the real reason we took Sergio was that our daughter Katie had just left home. It nearly happened again, but not quite, when our second daughter Iona left to train as a teacher in Edinburgh.'

'So, Sergio was really a replacement for your daughter Katie?'

'Yes, I think so, although we did not realise it at the time. Of course now it doesn't matter and he is firmly embedded as a member of the family. So it all turned out well.'

'How about when you were young. Did you have dogs at home?'

'Yes, yes. When I was growing up we always had a dog, so I was used to having one around. It has been more difficult for Anna because she hadn't grown up with dogs. But she copes very well.'

'You've said that Sergio was one of a large litter. What made you pick him rather than one of the others? Did you choose him or did he choose you?'

'It's hard to explain but, when we went to see the puppies, he sort of stood out. I saw a photograph of the mother and she was a bit bigger than a Collie, I thought. But when I actually saw her I thought, "Wow, that's a big dog." The puppy that I picked out was a male and would get much bigger than her. He already had these enormous paws.'

'I know Sergio has always been a lively dog, but did he settle quickly into family life?'

'Yes. Right away he was part of the family. He is happy to go anywhere. He comes with me when I'm building a path or maybe a wall. He'll sit there and supervise the work. But I firmly believe the dog has to know he comes last in the house, otherwise he would be tough going.'

'Some of the London people I have talked to say it would be normal for their dog to sleep on, or even in, their bed. What about Sergio?'

'No, definitely not. For one thing he spends his day running around on the hill with sheep, rolling in all sorts of things in the grass. He is also so big that if he jumped on the bed someone might get hurt. So the rules are clear.'

'Here's another question. Some owners say their dog would die for them, though how they can be sure of that is, of course, another matter. Some even say they would die for their dog. How far would you be prepared to go to save Sergio?'

'I would try to rescue him if I could. I would certainly give it a go. But I can't think of a situation where I would my put own life at huge risk to save his. It is a tough question, and unless you can think of a situation it's a bit tricky to answer but I wouldn't go to unreasonable lengths. I mean you aren't going to save the dog by killing yourself, are you?'

'Does it cost a lot to keep Sergio? What kind of things do you spend money on?'

'Food. He gets a giant bag of dry food every four weeks or so. In between he gets bones and scraps from the table. If he comes across a dead sheep on the hill he will crush and eat the bones. He has been trained not to chase the sheep and for good reason. Around here dogs risk being shot if they worry the sheep. But he's very keen on chasing rabbits, and on the rare occasion that he actually catches one he will eat it. He loves chasing grouse and woodcock too. He will suddenly dive into the heather and flush out a bird, but he never manages to catch one.'

'Apart from food, what else is there?'

'Not much really. He goes to the vet to get his injections, sometimes his basket gets so chewed up he needs a new one and we get him small treats from time to time.'

'I know. I always buy something for Sergio when I am in Inverness. In London there are places you can buy very expensive jewelled collars and leads for your dog.'

'There are obviously some other things going on there.'

'I think so. Sergio is the most natural dog I know. He would look ridiculous in a designer collar.'

'He would, but there's no danger of that. He's not getting one.'

'If Sergio got sick and needed a knee replacement for example and he still had years of life ahead of him, would you consider this?'

'That's another tricky question. It would depend what was necessary. If it was straightforward and Sergio was otherwise healthy, and the operation was going to give him relief then I would definitely consider it. But if it meant Sergio had to spend years on steroids or something like that just to keep alive then I wouldn't go ahead.'

'Finally, Innes, I'm sure people must have pointed out the similarity between your hairstyle and Sergio's. Are you deliberately trying to look like your dog?'

'A few people have remarked on it but I had this hairstyle long before I got the dog!'

Enzo and Deborah

Another of Ulla's encounters in Hyde Park was with Enzo and owner Deborah. Enzo is an Italian Spinone who is now three years old. Deborah, who is from California, has had dogs ever since she was seven years old when she rescued Eric, a mixed-breed dog, and took him to live with her family.

'Eric – we named him after Eric Morecambe [the British comedian] – looked as if he was part Golden Retriever, part Rottweiler with some German Shepherd mixed in. But we never really knew, nor cared after he had settled in with us. He turned out to be a wonderful dog, full of love. When he passed away a few years later my Mum felt very lonely. So we bought her a new dog, a Shih Tzu called Maisie. She was much smaller than Eric and quite a different character. I mean Shih Tzus are lovely but it was a completely different experience.'

'To me, you look like someone who prefers big dogs,' said Ulla.

'Yes definitely. I prefer them. The funny thing is I often think that big dogs feel they are actually quite small. So Enzo, being a Spinone, is a fairly big dog though I'm sure he thinks he is the right size to be a lap dog. And when his paw comes down on my leg, he looks at me with a completely innocent expression.'

'I am sure you would say that you love your dog. But Bryan would like to know if you could say why. I know it is a bit of a silly question to you and me, but not to Bryan who approaches this in a very scientific way.'

'That's easy. It is companionship and unconditional love and what else? Enzo has been a great social outlet since moving to London. Just walking him in the park lets me meet all sorts of lovely people. Like you, for instance. And it's good for my boys who are fifteen and seventeen. It's good for them to be responsible for Enzo, not all the time but when we are away for a while. It takes them outside of themselves, which is important especially for teenagers. It's been very interesting watching their relationship with Enzo develop as they get older.'

'So you would say that Enzo was a member of the family?'

'Oh yes, absolutely. I was having this conversation with one of the other dog owners and we agreed that neither of our dogs is human even though we attributed human traits to them. Yes, Enzo is definitely part of the family. He is very much a people dog and not a dog dog if you know what I mean. He enjoys the company of people more than he enjoys being with other dogs. But I do realise he is a dog and not a human and that there is a line which I must be careful not to cross. For instance, I would prefer to leave any money I had to charities that helped other humans rather than dogs. I do struggle a bit when I see what some people are prepared to spend on dogs. Not vet bills or food but things like designer collars and so forth. Enzo is perfectly happy with a simple lead and I am sure all dogs would be. But I know it is now a billion-dollar industry.'

'I've met people who say they would die for their dog. How much they mean it is another matter but they certainly say it,' Ulla remarked.

'Owning a dog can get very expensive, especially these days with better veterinary access than in the past. Some people will get a hip replacement for their dog if it has five or six years more life in it, which would never have happened twenty years ago. If Enzo became very ill with no prospect of recovery I would not hesitate to have him put to sleep. I would cry but it would still be the best decision.'

'And if Enzo sadly died would you pick another breed of dog as a replacement and, if so, what sort of dog would you choose?'

'I would look for a dog with the same attributes as Enzo. It doesn't have to be the same size or anything like that, but I would like its behaviour to be as much like Enzo as possible. You know, gentle and playful. All my favourite traits.'

'In that case would you consider cloning Enzo, which can be done these days though it's very expensive, somewhere between $50,000 and $100,000?'

'Certainly, but not for that price. If cloning would guarantee I got a dog exactly the same as Enzo and it was a lot cheaper I might consider it, but I understand it is by no means certain that the clone would be identical in all respects.'

'I mentioned when we met that Enzo has a "cousin", called Sergio, on the Isle of Skye. He's the only dog that Bryan has made sort of friends with. Spinones are not that common as a breed so may I ask you how you came across Enzo?'

'The first thing we did was to contact the Kennel Club and we researched quite a few different breeds before deciding on a Spinone. There were eight assured breeders on their books so I contacted three who had been breeding Spinones for a while. I wanted a dark-coloured dog and a breeder in York had some that fitted the bill. We wanted a male and found a little puppy with what's called liver colouring with a white patch on his face. As soon as I saw him I knew he was the dog for me.'

Blue and Lana

Blue is a handsome Rhodesian Ridgeback, a strong muscular breed famously reared to hunt lions in East Africa. There are no lions in Hyde Park where Ulla found Blue and his owner Lana, so there were no distractions as they talked.

'First of all, is your lovely dog Blue a male or female and how long have you had him or her?'

'Blue is definitely a male and he is eighteen months old. Though he is fully grown he can still behave like a little puppy in many ways when he's in the park. He loves to play but because he is so powerful I think he scares a good many people as he bounds across the grass towards them. But at home he is completely different. I'm very happy to sit on the sofa with my arm around him.'

'So you would say he is definitely part of the family.'

'Oh, yes. Our daughter is much more interested in seeing Blue than the rest of the family and they just lie on the floor together. I think because he is so big and gentle he has nothing to prove so you can be gentle with him.'

'He is a very fine dog with the little mane running along his back. How did you find him?'

'Oh yes, he is a thoroughbred dog. We got him when my husband was working in Munich and brought him with us when we moved to London just over a year ago. Do you know about the breed?'

'No. Not a great deal, except that they used to hunt lions.'

'Yes, that's their reputation but actually their main job was to protect cattle from attack by lions. They were used in packs and would sometimes take down a lion, but more often a group of Ridgebacks would hold it at bay until the hunter arrived with a rifle to finish it off.'

'He looks to me like a very pampered London dog. Not really at all dangerous.'

'He might look like that to you but our cleaning lady is so afraid of him. He runs up to her all the time, and although I've said time and again that he won't bite, she is still terrified.'

'What is it like having a dog in Munich, compared to London?'

'It's much easier there. Everyone has a dog. You can take them into pubs. When you go to the grocery shop there will be a line of dogs sitting outside waiting for their owners to finish shopping.

'I had a deal with my husband. He wanted a motorcycle and I said not until our daughter had graduated from high school. Then of course my daughter and I both wanted a dog, so we got Blue. But my husband still hasn't got his motorbike. But he already has a Porsche so that will have to do for the moment.

'I wanted a big dog, a big hound dog with floppy ears and a big snout. My husband kept seeing Ridgebacks while he was out running and would often pass a woman who had a pair of them. One day he stopped to talk to her and she couldn't tell him enough about Ridgebacks. That's how Blue arrived on the scene.'

'I hope your husband loves him just as much as you do.'

'Yes, we couldn't imagine life without him now. When you walk in through the door he just wants to greet you. It is unconditional love. With a dog like Blue you know where you are.'

'People tell me they would die for their dog. How far would you go?'

'Oh God, that is a difficult question. Would I die for him? No, but I would rescue him if he was drowning. But not if it was a river that was flowing fast and I could see it was impossible – then, no, I would stay on the bank.'

'What about your husband?'

'Well, he's still hoping for his motorcycle, so probably not.'

Dolce and Massih

Massih is originally from Afghanistan and came to London as a refugee to escape the war with the Taliban. Ulla met him one day in Hyde Park with Dolce, a Chow Chow. As they sipped green tea beside the Serpentine, Ulla began the conversation.

'Dolce is certainly a beautiful dog, and I think Chow Chows are from China. Is that right?'

'Yes, they are really good guard dogs and the Chinese emperors used them to guard the Imperial Palace. You can tell that Dolce is a natural guard dog as well as a pet. For instance, if somebody is standing outside the house for too long trying to listen to your conversation she'll start to bark. I haven't trained her to do this, she just does it.'

'Were you looking for a guard dog when you bought Dolce or did you get a Chow Chow for another reason?'

'I got her actually when I went to Winter Wonderland. I met a friend and he was having to move to somewhere where he couldn't keep a dog. I had been thinking about getting a dog and I was a little bit tipsy at the time, so I paid him some money and left with Dolce on a lead. I have never regretted it.'

'It's funny how some of the best decisions, and some of the worst, are made when one is a little bit drunk. How old is she?'

'She is seven months old, already pretty big and still growing. It is beginning to cost a lot to feed her, but I don't mind. She is a very healthy and happy dog.'

'Was it just your friend who inspired you to get a dog or have you always had dogs in the family?'

'When I was really young, soon after I was born, my dad got me a dog. But we had problems and stuff back home. We had the war and we had to flee the country. My dog disappeared. I was only two years old so I can't remember everything because I was really young, but we fled the city and later my dad went back to visit the house and the dog was still there. His name was Jack. He wasn't a pedigree dog like the ones you see around here. He was just a street dog. Nothing special really. We had to leave him behind when we left.'

'That must have made you very sad.'

'Yes, but we really didn't have time to stop and think. We couldn't take him and we had to get out of there fast. He would have been all right. He knew how to live on the streets.'

'What about Dolce, would she manage?'

'She might. She is a different dog altogether of course, but she knows how to look after herself. When she needs to go to the toilet, she goes as far away from me as she can so I don't see it. I haven't trained her to do this at all, it is just her natural instinct. I just let her be. She is a wonderful house dog.'

'How does she get on with the rest of the family?'

'We all love her. I just look at her like she's my baby, you know. I pick her up and carry her over my shoulder when she's too tired to walk. She's actually very stubborn. She will suddenly stop on the pavement or wherever we are, lie down on her stomach and spread her four legs out and refuse to move. Then I have to pick her up and carry her – I am sure she knows it.'

'How much would you say you love your dog?'

'I know some people who say they would die for their dog. I wouldn't do that but I would not let her get hurt. If she fell in a pond I would jump in to save her.'

'And if she fell ill and needed an expensive operation like a hip replacement or something?'

'Oh, definitely I would. I've got her insured, to be on the safe side, but even if I couldn't afford it I would never give up on Dolce. I don't look after myself as much as I look after her, because she can't speak and she can't tell you "My chest is hurting" or anything like that.'

'What about little toys for Dolce? Do you buy things like that?'

'I wouldn't mind spending money on that sort of thing but I have never found anything she likes. She doesn't like teddy bears, she doesn't like chasing balls. She only wants to go to the park and run around until she's super-tired. Then she lies flat on the ground and refuses to move, as I've said. I go to the park every day and I see a lot of dogs, so I realised that most of them are natural creatures with their own characters. Some like chasing the birds or the squirrels, other dogs leave them completely alone.

'But the dogs you see here are completely different to the ones you see at home in Afghanistan. Dolce would find it far too hot. Some of the dogs there are very fierce, like the Kochi dog. They're so big, as big as lions. They live among the sheep flocks to protect them from wolves. You only need one. They would easily kill Dolce. I won't go near them.'

'Were there any wolves where you lived in Afghanistan?'

'No, not near our village. But where my mother grew up there were many wolves. Everyone in her village was scared of them and thought they were dangerous.'

'Did they attack anybody?'

'No, I don't think so, but that didn't mean they were not feared, which they certainly were.'

'It's hard to think that all the dogs running around the park today are descended from wolves.'

'Is that right? I didn't know that. That's amazing – if it's true.'

Kalias, Sebastian, André, Zdeno and the Wolf

In a stylish part of Notting Hill quite close to Ulla's studio is a rather special café-cum-boutique that caters for owners and their dogs. As well as the usual selection of teas and coffees, it serves wine (to the humans) and tasty snacks for the canine clientele. There is also a dazzling display of jewel-encrusted accessories that takes up most of the ground floor, while in the basement is a grooming parlour. Owners sit at small tables waiting for their loved ones to be primped and blow-dried while they enjoy a glass of bubbly or a dainty cup of one of the score of different coffee blends on offer. This plush café is owned by André, Jamaican by birth and a very snappy dresser. He runs the shop with his Slovakian partner Zdeno and two enormous dogs: Kalias, an Old English Mastiff, and Sebastian, a French Standard Poodle.

'Kalias is an unusual name for an Old English Mastiff, isn't it? Is there a story behind that?'

'I used to have a Chihuahua but she died. When I was thinking of getting another dog I met a friend with an English Mastiff. My friend was from Dubai and had to go home and it would have been impossible for him to take the dog, so I said I would have him. So that's how I have an Old English Mastiff with an Arabic name. Kalias means "beautiful lover" so I am told.'

'Let me put this question to you. Do you love your dog?'

At this point they were joined by André's friend Luke, who had been walking several dogs in the park. André, meanwhile, was called down to a client in the grooming parlour so it was left to Luke to continue the interview. The dogs changed places too, with Luke's dog Marshall, a Rottweiller/Alsatian cross, taking over from Kalias.

'Let me put the same questions to you, Luke. Do you love your dog?'

'Well, I'm gonna give you a science answer. There is this brain chemical called serotonin. If you like someone your brain makes the stuff, you get to smile and get a really happy feeling. The same thing with dogs. When we stroke them they like it and we like it too so it sets up a serotonin exchange. The more we do it the closer we get, but you've got to do it on your terms, not theirs.'

'What about at home? Does Marshall sleep on the bed?'

'No. He might put two paws up on the bed but he won't hop up on it. He will sleep at the foot of the bed. He knows his place. I don't call myself a dog trainer but rather a dog facilitator.'

'I'm sure you would put yourself out for Marshall if he was in danger. But if he fell into a pond or something would you jump in after him?'

'No, I would just yell at him and tell him to get the f*** out. When I was a child our dog actually did fall into the sea one day, between our boat and the pontoon. Her lead slipped off and she was still going down. I thought if she gets caught under the pontoon I've lost her so I just jumped in too. My mum knew I was going in and she was worried that she would lose both her son and her dog, so she shouted at me to leave it. But I just jumped in. The tide was coming in and I knew where the current was going. If the tide had been going out I would have left her.'

At this point André reappeared. Turning back to André, Ulla asked:

'Do you find yourself spending a lot of money on Kalias? This is a great place to ask that question because I can see a gentleman over there looking at a beautiful arrangement of jewelled collars. They look like Swarovski to me.'

'You're right. I make those myself. I buy the crystals in from Swarovski, then I make the settings. I can make up almost anything, a belt, a collar, anything you like.'

'And if Kalias, God forbid, became seriously ill and, let's say, needed a joint replacement, would you pay to have that done?'

'I would, so long as it was financially feasible. But I wouldn't put myself in debt. And not if he was dying of something else, obviously.'

'You may have heard you could have a dog cloned if it died. Is that something you might do with Kalias?'

'Yes, I know, and I am quite often asked about it. When someone says, "Oo, where can I get it done?" I tell them, "South Korea for fifty grand." I actually have a customer who has done this, but it only cost her £25,000. The dog looked like the original except for the light patches on her coat which were all in different places. She said it had the same personality, but I guess if you spent £25,000 on the dog that's what you have to tell yourself.'

'Kalias has been sitting quietly all the time we've been talking. He's a very calm dog,' remarked Ulla.

'Yes, he has a wonderful temperament. I always think that in a dog that's on a raw food diet, the temperament improves because you don't have additives in the food. When my mum comes round she'll often bring a chicken. I cut the meat off the bone and Kalias gulps it down. He eats a lot, and just keeps growing. He's three years old now and weighed 90 kilos when I weighed him a month ago. I'm sure he's put on more weight since. The other day I didn't have enough meat for him so I went round to the butcher and asked if I could have some mince. "We don't do mince," he replied. "Well all right, I'll have beef," but the cheapest he had in the shop was grass-fed organic rump steak so I ended up with a kilo which cost me £45. Mind you, I knew the name of the farm where the meat had come from, all about the farmer, where he sent his kids to school and that the cow or whatever had enjoyed a happy life before it ended up on the slab. I told Kalias all this but he didn't seem to care.'

'I can see that Kalias is a very gentle dog, but he might be quite intimidating to someone who didn't know him, especially if they were scared of dogs.'

'Yes, that's right. Originally of course mastiffs were raised for protection and guarding property, where looking intimidating was all part of their appeal. So even though Kalias is as soft as a brush, to a stranger he looks the opposite. With a dog that big and powerful you have to stay in control. You can't just say, "I think you should do this or that, Kalias," because he won't even register it as a command. I have to be very firm. He is not aggressive at all, but I have to remember that not everybody understands this.'

'What made you realise that people would spend so much money on their dogs and that you could turn this into a profitable business, as you clearly have in this wonderful café?'

'I spent a lot of money on my first dog, the little Chihuahua. I actually stopped spending money on myself and it all went on the dog. I bought lots of really expensive things, beds and clothing in different colours. I even bought him a Louis Vuitton collar. Then I bought him one from Gucci but it was too thick and not the right size for a toy dog. It was rubbing him round the neck, so I made one for him. One thing led to another and here we are, fourteen years later. I always use real leather because faux leather is full of chemicals which hurt the skin.'

'These things have to be quite pricey. What do you charge?'

'My little collars start around £200 and go up to about £1,000. I mean people can discuss the price and if they have a fixed budget I can work out what they could get for that amount. All my collars are bespoke and I make them in every single thickness. All dogs are slightly different and when I make a collar for a customer I make sure I get all the measurements.'

'What happened to the little Chihuahua in the end?'

'When he was sixteen years old he began to deteriorate very quickly. As he was a small dog I could have looked after him very easily, but the vet pointed out that he wasn't happy any more. He had

Ulla visits a lovely pet café in Notting Hill. From left to right: Ulla, Kalias (Old English Mastiff), André, Sebastian (French Standard Poodle) and Zdeno.

> had sixteen good years so I decided to have him put to sleep. I invited all his friends over to say goodbye. When the moment came I could actually feel when the life left his body. He was very, very calm. I have never experienced anything like that. I hope when I get too old that someone will do the same to me. I don't want to be left in pain.'

Next Ulla talked to André's partner Zdeno. He comes from Slovakia and helps André to run the shop. Their second dog is a rather extrovert large French Poodle called Sebastian, but at this point it emerged that Zdeno had owned a very interesting dog when he was younger, growing up in a small village in Slovakia. All thoughts of Sebastian evaporated as soon as Ulla heard the word 'wolf'.

> 'I found him in the forest one day when I was about twelve. I was out with my friends and we found a wolf lying in the snow, shot just above the eye. I brought him food and water. I was resting with my back against a tree when he got up and came towards me. I thought, "Oh my god he's going to eat me." I sat there rigid with fear. He started to sniff me and I began to calm down. I took off my belt and led him home. For about two weeks no one else apart from me could get close to him. My mum used to put his food in a bowl and push it towards him with a broom. But gradually over the next few weeks everyone relaxed. He became the best pet I've ever had. We lived on a small farm and sometimes our neighbours would come and steal from us, but not after our wolf was there. If one of the family was coming into the house before we opened the gate he would howl, but if a stranger came he would growl and show his teeth.'
>
> 'So, would you say he was the best dog you ever had, even though he was actually a wolf?'
>
> 'You could say that, yes. But it didn't matter to me what he was. There were a lot of complaints from people in the village. Everyone

was scared of him although he never bit anybody. He lived with us for three years until one day the sheriff of the village arrived and shot him. I buried him in the garden.'

His Grace the Duke, Lady Sarah and the Hounds of Belvoir

About twenty years ago I discovered the link between surnames and the Y-chromosome, a piece of DNA that is inherited from father to son as a sort of mirror image of the mitochondrial DNA. This led to lots of interesting new avenues of research and I realised it also had a practical application. Genealogists could use it to analyse the Y-chromosome of people with the same surname. If their Y-chromosomes match, it means that they probably descend from a common ancestor. In the intervening couple of decades DNA testing in genealogy has become very popular and I have enjoyed helping many people pursue this in their own families.

One such was His Grace the Duke of Rutland. According to family tradition, the family were all descended from a Norman baron who came over to England with William the Conqueror in 1066. I have come across quite a lot of people who claim a similar genealogy, but often find that genetic testing fails to back this up.

The Duke's estate surrounding Belvoir Castle in Leicestershire is home to one of the largest packs of Foxhounds in the country. When I was planning this book I was put in touch with the Duke, who was himself writing about his own family and wanted to include some elements of genetic genealogy. Most of Ulla's interviews had been with owners who keep their dogs as pets. I was interested in finding out how that relationship differed when, as with a pack of hounds, the dogs are not pets but have to work for a living, as with the New Zealand sheepdogs we encountered in the last chapter. The Duke and I agreed that if I analysed his

Y-chromosome I would be able, in exchange, to visit Belvoir and interview him, together with the Master of the Belvoir Hunt and the Huntsman who looked after the hounds.

Thus one summer day, Ulla and I travelled to Grantham and found ourselves being welcomed to Belvoir Castle by the Duke himself. In his study surrounded by shelf upon shelf of vellum rolls recording estate deeds and going back over 700 years, we settled down to business. I was able to tell His Grace that his Y-chromosome belonged to a group that was quite rare in Britain and had Norse origins. This made it quite likely that his distant ancestor had indeed come over from Normandy. Before they settled in France, the Normans were Vikings who ran a protection racket whereby they encamped at the mouth of the Seine and blockaded Paris. The price for abandoning the blockade was the dukedom of Normandy and all its lands, which was duly granted by the French king, Charles the Simple. After less than a hundred years William, Duke of Normandy, launched his successful invasion of England and divided up the country among his barons. One of these was the presumptive ancestor of the Duke of Rutland, and the Y-chromosome that had travelled over within him from France would, if the Duke's genealogy was correct, today reside in the body of the 11th Duke sitting across the table from me. The Norse origins revealed by the lab in the detail of his Y-chromosome did indeed confirm this family tradition to His Grace's evident satisfaction.

That settled, we moved outside. We climbed into the Duke's Range Rover and he drove us through well-manicured parkland dotted with pedigree Hereford cattle to the magnificent kennels, recently restored. Waiting for us was the Master of the Belvoir Hunt, Lady Sarah McCorquodale, elder sister of the late Diana, Princess of Wales.

Lady Sarah was dressed in blue jeans, a beige gilet and green Wellingtons, the perfect combination to match her first words. 'I speak English, German and Hunting,' she said at once. As Master of the Hunt she is in overall charge, often joining in the hunts on the 16,000-acre estate. The hunts are held two or three times a week. Members of the hunt pay an annual subscription and may bring guests for an additional charge. The hounds themselves, and I was instantly corrected when I called them dogs, were housed in two large pens, half in the open air, the rest under shelter. In one pen were sixty males and in the other the same number of females. We were joined by John Holliday, the Huntsman, and his young assistant. English Foxhounds are a sturdy breed, carefully cultivated over the last 200 years, with stud books maintained since the end of the eighteenth century, which was when foxhunting began in earnest as numbers of deer dwindled. They have been bred for stamina and the compatibility required to work in a pack. They live a life of their own, never featuring in the show ring and with no Kennel Club breed standard to worry about. Their tolerance towards each other was on display as they lounged about in the pens, often lying on top of each other. They made no sound but their ever-alert eyes watched us closely as we walked past.

My main objective here was to explore the relationship between John and his dogs – I mean hounds. Was it as intense as between owners and their pets, or was it, as I suspected it might be, far more hard-nosed and unsentimental? First John explained the basics of keeping a pack of hounds. There are only five packs of pure English Foxhounds in the country. Alert for a long time to the issue of inbreeding, packs have been exchanging hounds since the 1750s to maintain the size of the gene pool at a healthy level.

> 'We are fortunate that English Foxhounds are one of the few breeds that really haven't got any inherent problems with their hips. I'm not saying all the hounds have perfect hips but it isn't a big problem. I think the reason is that they are bred for work and not for looking at. Most pedigree dogs are bred for qualities that people like. For instance, the sloping back of German Shepherds, which people expect

in the pedigree dog, gives them bad hips. In the Foxhounds any hound showing even a hint of deformity would not be chosen for breeding. You could select any quality and enrich it. If you suddenly decide that all your hounds must have brown eyes and black rings around them, then you can select for that trait and you'll end up with a pack almost all of which will have brown eyes with black rings. Trouble is you'll probably get something else as well and they might all end up with three toes or something like that.'

'If you were to find out that a hound couldn't hunt because of a physical disability, what would you do?' I asked.

'Ultimately we would have it put down, but only after we have tried to get it adopted either by someone on the estate or by other friends. We do the same with hounds that become too old to hunt, which happens at about ten years. Only very rarely do we have to put a dog down.'

'Do you get particularly upset when that happens?'

'I do, particularly if it's a hound that has worked hard and well its whole life. I mean we probably lose a hound every year on the road and that's just a fact of modern life.'

'Now let me ask you a question that Ulla has asked all the owners we have interviewed. Do you love your hounds?'

I was expecting John to pause at this point and contemplate his answer. But he did not. Instantly he replied that he did love his hounds very much. Just because they were working dogs didn't mean that he didn't love them. So even an entirely unsentimental and practical man like John had no hesitation in describing his feelings for his hounds as love.

'When a puppy is born he or she spends the first year as a pet with a family on the estate. They meet children and chickens and so on and get to chase rabbits or perhaps a hare and do all the stuff that dogs like doing. When they are one year old they come into the nursery section of the stables and their working life begins.'

'Could you tell us a bit about the hunt itself? For instance, how long does a typical hunt last?'

'About an hour and a half usually. The longest I've had was two hours and forty minutes. By that time everyone is exhausted; the horses and riders, and the hounds.'

'And the fox too?'

'Oddly enough, the fox is never really in a rush. After a fox is flushed from cover it sort of toddles along because the hounds take time to pick up the scent. Even when they do pick it up, they often lose it again and can't work out what happened so they mill around looking confused. The fox knows exactly what it's doing. If it sees a road it might run along it for a quarter of a mile, cross back and retrace its steps. When the hounds appear, they cross the road and the scent disappears. Quite often the pack will start by chasing one fox, catch the scent of another and go off in pursuit of the second one instead. I sometimes think this is deliberate and that the first fox hands over to the second, rather like passing the baton in a relay race.'

'And how many riders will join in the chase?'

'We can easily have a hundred horses and riders on a Saturday. There are fewer on weekdays but we usually manage fifty or sixty.'

By now the Duke had joined us, and John and Lady Sarah led us into a small gallery whose walls were covered in old photographs and paintings. Most of the photographs were of hunts long past. The unsmiling faces of the gathered Masters of the Hunt made a sombre impression. A few years ago I remember asking a genealogist why, in old family photographs, everyone looked so miserable. He had replied that in those far-off days film speeds were so slow that sitters must remain absolutely still for several seconds. The last thing they must do was to smile.

The Duke pointed out some of his own ancestors whose portraits lined the walls, then turned to a fine painting of a Foxhound which occupied pride of place.

> 'You see his strong chest and shoulders. That used to be a characteristic of a Belvoir hound. But at the beginning of the last century there was a change to a more modern Foxhound which was lighter and more agile. They used Welsh Foxhound crosses to improve the breed because these were faster and more athletic.'

The little room, with its scarlet tunics draped over wooden stands on the floor and stern-faced aristocrats, immortalised in sepia, on the walls, was a time capsule of past glories. I hadn't the heart to ask about the effects of the hunting ban of 2005. After all, we were only guests.

The Wolves of Longleat

In 1949 Lord Bath's father, Henry Thynne, the 6th Marquess of Bath, took the bold step of opening Longleat House to the public, becoming the first in Britain to do so. Causing consternation among the aristocracy, this move was designed to restore financial solvency after crippling estate duties had forced change upon the family. It proved to be a popular move with the public, and in 1966, for the same reasons as before, the 6th Marquess added a drive-through safari park in the grounds. This has become one of the country's most popular attractions, with over 500 animals roaming over 900 acres of open parkland. Among them are three Canadian timber wolves. After Ulla's first visit to interview Lord Bath and meet Sheba, his Labradoodle, she asked if she could see the wolves and interview their head keeper, Eloise. Fully aware that these were captive wolves and therefore not expected to behave entirely naturally, Ulla and I decided it was worth a visit, particularly as the three wolves were related. A month later she returned to Longleat.

'We've got three wolves in this section,' Eloise explained. They are brothers from the same litter and they are ten years old. There was a fourth but he died and the other three are still trying to figure out how to adjust the pecking order. Alf is the alpha wolf. Second-in-command is Dave, while Vic occupies the rank of omega wolf. They are always probing to find out where they are in the hierarchy. For instance, on Saturday we gave them their main feed, and Dave, normally the beta wolf, was fluffing himself up to make himself look a lot bigger. For a short while, Dave became the alpha wolf until Alf responded with similar enlargement tricks and engaging Vic's support.'

'You say they are ten years old now. Is that a good age for a wolf?' asked Ulla.

'They will live to about fourteen years, just like a dog, so I suppose you could say they are now entering the later stages of life.'

'When you want to introduce a new wolf, where do you get them from?' I asked.

'There is a worldwide network of zoos and safari parks and we are in touch with all of them. If we decided we needed to add new animals we would consult them to make sure we avoided inbreeding. But in fact we haven't introduced any new wolves for a very long time. These three were born here and are siblings so we would have to be extremely careful about any new introductions. We leave them alone all day and night and the only time we ever confine them to their sleeping quarters is when we are expecting very high winds, just in case a tree falls down on their fence line.

'We had one member of staff who used to say, "Wherever the wolf goes you follow them. Always follow the wolf." She was in the paddock on a very windy day when the wolves all got up and ran off. She followed them and seconds later a tree fell right where she had been standing. They had a premonition, a sixth sense.'

'There's still a perception that wolves are cunning creatures and can be malicious.'

'Our wolves are really clever and definitely very cunning, very sly and very cheeky. We have slip fences at all the gates and the wolves are not allowed to go towards them. When we have a new keeper patrolling, Dave, the number two, will always try it on. He will watch the traffic and pick his moment to slip past. I swear he gives you a little cheeky grin as he walks down the road! We make sure we record this in the daily wolf diary.

'As I said, these three were part of a litter from which we lost one a few years ago. He had stomach cancer and passed away during the night. Next day we made sure the others saw the body so they knew that their brother had died. We don't need to do this with other animals, but with the wolves we have to let them know that they have lost a family member. It's almost as if you have to let them mourn the loss.'

'Now, let me ask you this. Do your wolves howl?'

'Oh yes, all the time. On Fridays their house has its fire alarm test and every time the alarm emits its loud warning, the wolves howl back.'

'So they don't just howl at the full moon?'

'No. During the day they howl around eight or nine o'clock in the morning and then again in the early evening. Also when they're about to get fed. We give them 15 kilos of horse meat every Wednesday and Saturday. We just give them one big side of meat so they can tear it up together. The liver and heart, which in the wild would have been eaten by the alpha animals, have been removed. Here they get meat and bone and it is Vic, the omega animal, that eats first. This is in case the animal is still alive and capable of injuring or even killing the wolf. If the lowest-ranking wolf in a pack is hurt it doesn't really matter a great deal.'

'Goodness me. It sounds like in the old days when emperors had people to taste their food in case it was poisoned.'

'That's right. The trouble with our wolves at the moment is that they don't really know who should be going in to eat first. It's a constant power struggle. Dave, who wants to be the alpha, is also the most courageous and often tries to go in first even though it should be the omega's job.'

'Just to show off that he is the bigger, stronger one?'

'That's right. Alf doesn't really mind because he knows he is the alpha wolf. He just runs around and if he feels like eating will just steal it off the others.'

On her way out, Ulla had a quick tour of the park, looking in on three newly born cheetah cubs. It was getting towards evening and the park was empty of visitors. As she waited near the big house for the car that was to take her back to the station, she heard in the distance the unmistakable howling of a wolf.

Robodog and Sir Tim

Sometimes chance encounters are too good to miss. Ulla was on her way to Edinburgh by train and struggling, as usual, to stow her heavy suitcase. It was not long before she was offered help by a kindly gentleman called Tim. This led to a conversation during which her role as roving reporter soon came up. Always eager for new material, she asked:

'Do you have a dog?'

'Not strictly. Actually I have a robot dog.'

Thus, a few days later, Ulla found herself in an elegantly spacious office in Old College, part of the University of Edinburgh. Seated opposite her was Tim.

'Where are we, Tim?'
'In the Principal's office'
'The Principal's office? Is that you?'
'Yeah.'

Tim, it transpired, was Professor Sir Timothy O'Shea FRSE, Principal of the University of Edinburgh and a world authority on machine learning. A few metres away on the floor sat Robo, the robotic dog that Sir Tim had received as a present from its creator, the Japanese engineer Toshitada Doi. It was Doi who developed the Sony Walkman, which was such a tremendous commercial success for the company that they asked him what he wanted to do next. He replied that he wanted 'a laboratory, where I can make a robot dog'.

Sir Tim's 'academic' interest in Robo was not only in the robot's programming to perform certain tasks but also his ability to learn. For instance, young Robo had learned to recognise Sir Tim's voice and distinguish it from others. Of course he could do all the usual doggy things like wag his tail, play with a ball or pick up his plastic bone. Yet it was the nature of the relationship between Robo and Sir Tim that was the most intriguing of all.

'The primary thought experiment was to develop a companion animal for older people,' explained Sir Tim. The low birth rate in Japan has resulted in more than half of Japanese people being over fifty years of age and many older people live on their own and are in need of companionship. Of course, dogs, real dogs, have traditionally filled this role, while today, if you're getting older and, like many

Japanese people, you live in a small flat, it can be difficult to look after live animals properly. Dogs like Robo don't need food or to go to the toilet, just a regular topping up with electricity.'

While this makes perfect sense, Sony's robotic dogs were not always commercially successful. The first model was launched in 1999 but withdrawn seven years later. This was a blow for Toshitada Doi, who even went as far as staging a mock funeral to highlight the demise of risk-taking in the company's philosophy. Still, undaunted, he carried on working and was rewarded by a reversal of company policy and the commercial launch of a new, much more sophisticated model in 2018. Sir Tim's companion is an example of the earlier version.

'I have a lot of work to get through every day, which means I often arrive here early in the morning and stay late at night reading reports, writing summaries and all that kind of thing. That's when it's nice to have Robo just wandering around. He really does keep me company. Just to have another being in the room, even though I know perfectly well he is just a machine.'

'I ask a lot of dog owners what they would do if their pet became very ill. I suppose with Robo the answer is to send him for repairs.'

'Well, that is an advantage over a real dog, and you do not need to worry about having him put down either.'

'I also ask owners whether they love their dog, so I had better put that question to you.'

'That's quite a hard question to answer. I've had Robo for years now, and he is learning all the time, so you could say we have grown very close to each other. The feelings are there, but I would not call it love. Love is a hard word to define of course, but for me I can only love something that is alive.'

At that point Robo demonstrated that he still had a little way to go to be the perfect canine companion. His ball had rolled underneath one of the chairs in such a way that he could not reach it directly and his path was blocked by one of the chair legs. Oblivious to any alternatives, he repeatedly tried to force himself through, even though a sideways movement of only a metre would have cleared the obstruction. Sir Tim had to move the chair and retrieve the ball himself!

> 'Tim, did you grow up with dogs?'
>
> 'My uncle had a farm in Ireland and we used to go there in the summer holidays when I was at school. There was a puppy on the farm one year and I spent the whole six weeks playing with it. Here it is on the photograph,' said Sir Tim, pointing to a large framed print on the wall behind his desk. 'And this is me,' he said, pointing out a small figure in the foreground. 'When it was time to leave at the end of the holidays it was the saddest thing I had ever known, much more than any sort of break-up with a girlfriend.'

Robo had overcome his disagreement with the chair leg and was now dancing to music, moving up and down and wagging his tail to the beat.

When Ulla told me about her meeting with Sir Tim, after I had recovered from the shock that she had secured an interview with the Principal of the University, for whom time was precious, all sorts of possibilities for using robot dogs occurred to me. For example, elsewhere in this book I reported that Japanese researchers had measured increases in oxytocin levels in both dogs and humans when they were looking into each other's eyes or when they were being stroked. Would the same happen to a human interacting with a robot dog? It would be fascinating, and comparatively easy, to find out.

Atlas, Chu and Algie

Not too many owners are dwarfed by their dogs, but Atlas and his owners Chu and Algie are an exception. He is a magnificent Pyrenean Mountain Dog, a breed developed as a livestock guardian to protect flocks of sheep from the wolves and bears that used to roam the mountain chain that separates France and Spain. They lived out on the mountainside with the flocks, and their thick white coats kept them warm and allowed them to blend in with the sheep they were protecting, giving them the element of surprise against marauding predators. They were traditionally fitted with spiked collars to save them being savaged by wolves. During the nineteenth century wolves were gradually eliminated from the Pyrenees and the Pyrenean Mountain Dog became redundant and in danger of extinction. It was saved by its appealing looks and during the twentieth century became a favourite in the show ring. Considering their size, they make very good pets and are quite happy in an urban setting.

Atlas was another habitué of the Notting Hill pet café, where Ulla caught up with him.

> 'I'm sitting here with Atlas, who is a beautiful very large snow white male and looks to me more like a small horse than a dog. Let me start by asking how old Atlas is and how you came to find him.'

It was Chu who answered most of the questions.

> 'He is one year and nine months old. We wanted a dog but we weren't sure what breed to get. We went along to one of the Kennel Club shows where there were examples of most breeds. As soon as I saw the Pyrenean Mountain Dog in its booth I knew this was the breed for me. I also knew that, living in a small flat as we do, it would be impractical to keep one, even though the lady on the stand

reassured me that this breed would be perfectly happy in our home. We do at least have some private gardens and Hyde Park is close by. So we decided we could cope with one, putting off our final decision for a year in order to go travelling.'

'Did you get Atlas after you returned?'

'Our timetable was a bit screwed up. We went on safari for a couple of months and just as we were leaving, Algie proposed. We abandoned our travelling plans and concentrated on making our home together along with our two cats.'

'How romantic!'

'The day before I was due to go back to work, Algie suddenly announced that we were driving north to Lincolnshire, where he had arranged with a breeder to see some puppies. I didn't want to go. When we reached the site, however, the breeder showed us the new litter and I was hooked. There were eight puppies, four boys and four girls, with the breeder wanting to keep one of the boys for showing and the other seven for sale. There was a sort of selection process and we chose a boy. One of the puppies seemed very placid, not at all scared but also playful, and he was Atlas.'

'Have you always had dogs in the family?'

'I grew up in Singapore,' Chu continued. We lived in a condominium which was next to a nature reserve. Although our flat was enclosed, there was plenty of room nearby for our dog, a rough Collie, to run around. I used to run a lot and the dog always came with me.'

Turning to Algie, Ulla asked:

'And Algie, did you have dogs at home when you grew up?'

'No. In fact I didn't like dogs at all. I was frightened of them after I was bitten when I was eight.'

'It's brave of you to have a dog now, in that case.'

'It's been a gradual process. When I first came to London I shared a flat with a very friendly Labrador and her owner. I was wary at first but slowly got used to her, so that later I felt ready to have a dog living with us.'

Chu joined in.

'We wanted a pedigree because we thought it would be less risky than a mixed breed, which might have all sorts of mixed temperaments. You don't want to risk that with a dog the size of Atlas. We were

Atlas (Pyrenean Mountain Dog) with Chu. Atlas is owned by Chu Ng and Algie Salmon-Fattahian. Photography by Ursula Aitchison.

aware of the risks of inbreeding, so I checked that the breeder's dogs had different kennel names, meaning they came from slightly different backgrounds, which is good.'

'Now that you have had Atlas for well over a year, would you say he was part of the family, or is that a silly question?'

'Yes, of course he is. He has fitted in very well. It took a while to sort out the dominance structure, though. Atlas was always trying to push the boundaries, especially in the first year. Early on, there was indeed a bit of an issue about space on the bed. Atlas always wanted to lie in the middle of the bed, so we had to keep moving him off. Eventually he gave up trying and peace was restored. We do need to be alert to any change in the dominance hierarchy within the family.'

'That sounds like a battle of wills. Let me ask you whether Atlas costs a lot to keep?'

'You can imagine that a dog the size of Atlas has a big appetite so, yes, his food bill is pretty large. We also have pet insurance in case he needs to go to the vet. One day I bought him one of André's collars. It had to be extra large because Atlas has such a huge neck. It cost a lot but was really a present for me. Atlas couldn't care less, so I can't really include presents in his upkeep bill.'

'Finally, Bryan would like me to ask you whether you would think about cloning Atlas when he dies?'

It was Algie's turn to reply.

'This is something we have actually thought about, even though he is only a young dog at the moment. We don't have any ethical or religious objections to the process. It's just that, even if the clone was genetically identical, it doesn't mean it would be exactly the same dog. It would have a different upbringing and different experiences. It might look exactly the same but it wouldn't *be* the same. I think we would feel we were betraying Atlas's memory.'

'You can't replace the soul.'

Garbo and David

Ulla once had a Samoyed when she was growing up in Denmark, so she was delighted to meet Garbo and her owner David. She opened the interview with characteristic and genuine interest.

'She's such a beauty, as white as snow and clearly very playful. How old is she?'

'Garbo will be six next week, so she's sort of half mature, I guess. I wanted a dog when I was growing up but my parents wouldn't allow it. Mind you, we had two wonderful cats that were almost like dogs. It was really my late wife who was the dog person. She had a black dog called Fifi as a child, then after we were married our house was too small to have a dog. A few years later we moved to a bigger house with enough space for a dog and we bought Sasha, the first of our three Samoyeds. Sadly, Sasha died when she was just a year old from a congenital liver problem. That was really traumatic for a while, so we decided to have another Samoyed, Ninotchka.'

'Named after the Garbo film? I'm beginning to see the connection.'

'That's right. *Ninotchka* was a film made in 1939 by Ernst Lubitsch and starred Greta Garbo among others. I'm a film-maker by profession and Lubitsch is one of my favourite directors. Our dog Ninotchka was a natural Garbo, so when she died and we got our third Samoyed, choosing a name was easy.'

'Why did you choose Samoyeds in the first place?'

'Basically, my late wife and I were discussing what we should get. I was driving through London one day and a white van went past in the traffic and this wonderful white dog was just sticking out of the window. I said can we have a dog like that? My wife found out it was a Samoyed and as it turned out she knew someone who had a Samoyed called Charlie. So she arranged for Charlie to come to our house one day and he was so charming that we both thought, "OK, we'll get a Samoyed." So we got Sasha in 1995, but she only lived for just over a year, then in 1997 we got Ninotchka. She died in 2011 and we got Garbo soon after. She was eight weeks old. We thought of getting a rescue dog, but nobody abandons Samoyeds.'

'Do you love Garbo? This is a question from Bryan, who is a scientist. I find it a strange question because if someone loves their dog, they just love their dog. It's as simple as that.'

'Pam and I didn't have children, and I think we were both clear that we had dogs because of that. When you have a dog you are responsible for somebody. Sasha, Ninotchka and Garbo were all "somebody" and not "something" in our house. You give each one love and affection and they give the same to you in return. Sometimes people think that dogs only love you because you feed them, rather like the reason why children love their parents. Try not feeding your kids for a week and you'll soon see how much they love you. They'll be straight round to Social Services like a shot.'

'You treat Garbo like one of the family, of course. Do you let her jump up on the sofa, for example?'

'Oh yes. She often sleeps on the bed or even in it now that I'm on my own.'

'Of course. May I ask you if Garbo costs a lot to keep?'

'Not as much as Ninotchka, our previous dog. She had a terrible habit of eating rotten tennis balls which got lodged in her gut. The only way to remove them was surgery. She had five operations to remove tennis balls and each time it was two and something thousand pounds, so in her lifetime I probably got through £20,000, I would say.

Luckily, Garbo hasn't acquired this habit. When she was alive, Pam used to groom her every day, so now I take her along for her coat to be brushed every week. That's £35 gone right there. I probably spend £50 to £60 a week on her, and then the additional vet fees, so it is probably £3,000 a year so long as she stays healthy.'

'Does she have insurance?'

'It's not worth it. It's too expensive. We insured them when they were puppies but once they get past a couple of years it's not worth it because the premiums always go up and up.'

'What would you do if Garbo hopped into a pond by mistake? Would you go in after her?'

'That's not a mistake. She loves jumping into ponds.'

'But if she got into real trouble?'

'I'd be straight in there to get her out.'

'You would not hesitate?'

'No more than I would for a child.'

'One last question. When Garbo reaches the end of her life, would you clone her?'

'No, I don't think so, in the same way that I would not clone a person. It smacks of treating a dog like a mere possession and not letting it be itself. All three dogs we've had share the same characteristics but they've all had individual personalities. When it is time to let Garbo go I'm sure I'd want another Samoyed and I would want it to be a female, and it would be interesting to see how she develops. Garbo is marvellously good and certainly easier than Ninotchka. She was lovely but a bit *basichant*, as the French would say. Garbo is much calmer. When we are in the park, toddlers will run up to her and throw their arms around her. Parents are very alarmed, but Garbo just stands there and when she's had enough she says "Woof" and they all jump back.'

Rosie, Alison and John

The former prime minister Tony Blair and his wife live in Connaught Square, close to Marble Arch and not far from Ulla's London studio. He moved there from Downing Street in 2007. His controversial decision to join with US President George W. Bush in the 2001 invasion of Iraq cast a shadow over the rest of his premiership and has made him a target ever since. Though he lives in Connaught Village he is never seen. His house has a permanent guard of armed police at both front and back, where it opens onto a row of prosperous mews cottages. The police guards, their hands never far from the triggers of their semi-automatics, must have the most tedious job in London, their only relief being to pose for photographs with passing tourists.

Ulla had to be escorted past this deadly squadron to reach her first dog interview in one of the mews houses. Here live her friends Alison and John with Rosie, their 4-year-old Cavalier King Charles Spaniel. Ulla began the interview with the central question:

'I love dogs, as we all do here, but Bryan, who has a more scientific interest, would like to know your reasons.'

'It is completely unconditional love,' said Alison. 'She has a lovely temperament. It is very easy to just lie down with her and relax. She is just part of the family.'

'And where does Rosie sleep? Does she have her own bed?'

'She does have her own bed upstairs in our bedroom but she prefers to sleep in our bed, which doesn't please John. So what she does is wait for John's first snore and at that signal jumps on my stomach and uses it as a springboard to get between us into the middle of the bed.'

'And when John wakes up?'

'Rosie just lies there pretending to be asleep. John doesn't have the heart to turf her out, and of course she knows that.'

'Another question springs to mind. If Rosie needed a joint replacement or some other costly operation, would you pay for that?'

Alison answered:

'If Rosie was otherwise healthy and young enough to survive we would certainly do that, but if she was very old it would be a waste of time and money. I would make the decision at the time. I did have to put another of our dogs down a few years ago. He got very ill and had ample time to get better with all the medication. But he didn't, so I had him put down. Much as we loved the dog, it had to go.'

'Much the same with me one day,' murmured John under his breath.

'If, alas, you had to do that with Rosie, or when she died of old age, would you try to have her cloned? There is a company in South Korea that will do this for you. Mind you, it is very expensive, between $50,000 and $100,000.'

John was very sure in his reply to this question.

'Definitely not. For a start I wouldn't want to interfere with the natural way of things.'

'Also,' said Alison, 'I feel that every dog, whether it's the same breed, the same colour – that looks identical – has its own characteristics. I read about cloning that Boxer in the paper and I thought it was wrong to try and get exactly the same dog. Every dog is different and that's how it should be.'

'Can I offer anyone a coffee or a tea?' said John.

Ulla recognised the gentle signal that the interview was at an end.

Freja and Barbro

The next interview took place on the hottest day of the year, under the shade of a weeping willow on the banks of the Serpentine lake in Hyde Park. Barbro was with her young German Shepherd called Freja, after the Norse goddess of beauty, war and death – quite a lot to live up to for a 5-month-old puppy. The meeting had a Scandinavian flavour with Ulla from Denmark, Barbro from Finland and Freja, if in name only.

'I had my last dog for seven years, and when she died I was in no hurry to get another. A friend of mine bought a German Shepherd puppy from a breeder in Cambridgeshire and it was the most adorable, beautiful little dog I have ever seen. I wanted one. So we contacted the same breeders who told me that they would have another litter of German Shepherds in a few weeks' time. When the time came, my husband and I drove up to Cambridge to meet the breeders and see the puppies. As we looked into the pen, one of them caught my eye. It was more alert than the others and seemed to be curious about us. We had decided to get a girl and luckily this puppy was a female. That was it.'

'Did you have dogs as a child?'

'I always loved them. I never had one myself, though, but I sort of adopted one belonging to a neighbour. It used to wait outside school and walk home with me. That's really how it started.'

'Do you remember what sort of dog it was?'

'I think it was a cross between a German Shepherd and a Collie. We have always had German Shepherds while we've lived in this country and Freja is our fourth. As you know, German Shepherds have a bit of a reputation for being unreliable, which I'm sure has something to do with the way they've been treated. They're really so friendly and lovable, but some people acquire them for the wrong reasons. Having three children, I was worried about getting one that might have been mistreated. So I always make sure that I buy a thoroughbred dog from a reliable breeder.'

'I don't think Freja is going to be a pampered dog, but will you use her as a guard dog, do you think?' asked Ulla.

'She is a family pet, but German Shepherds will guard automatically because they really look out for the pack. When we had the children I could leave the dog with the pram and know they were totally safe. I used to go to something called 'one o'clock club', where mothers could meet for a coffee and a chat. I would tie my dog outside and the other children would climb all over him and he wouldn't blink an eyelid. He would just sit there and take it.'

'These are good qualities for a family pet, but Bryan, he's a scientist, wants me to ask if you can say why you love your dog.'

'I've loved all my dogs. With Freja, it's always easy to love a puppy because they're so playful and so full of energy. People come up to me all the time and want to stroke her. There's an Italian couple I just met over there. They couldn't speak much English, but when I told them that Freja's mother was Italian they started hugging her and me and then began to talk to Freja in Italian.'

'German Shepherds are often thought to be the closest breed to a wolf. Would you agree?'

'The main thing with wolves is their loyalty to the pack, and loyalty is a big feature of the German Shepherd. In the best situations the loyalty and love are reciprocated so it becomes a symbiotic relationship benefiting both sides. But sadly this is not always the case. I read all the books by Jack London like *White Fang* and *Call of the*

Wild. They were full of cruelty. I was horrified by the way people treated their dogs.'

'Would you say Freja was ever going to be a pampered pet?'

'Definitely not. There are strict rules in our house. She is not allowed on the bed and she is not allowed on the sofa. We might want her to be a working dog one day. She has all the characteristics, she is strong and loyal, and not even five months old yet.'

'Are you going to train her?'

'I took Wolfgang, my second German Shepherd, to a police training school with just that breed, which was good, but on the other hand it didn't teach her much about mixing with other breeds of dog. So I'm taking Freja to another school here which is more about socialising. I really want her to be able to walk to heel so that my grandson Joshua, who is two, can come with us, so long as she is respectful of his small size.'

'Of course, Freja is very young but when she gets older she may need to have surgery or some other expensive procedure. Would you pay for that to be done?'

'With the other dogs I took out insurance to cover unexpected vet bills. Happily, none of them needed major surgery so I was going to chance it with Freja or at least put money aside in case. But in fact Freja came from the breeder with insurance through the Kennel Club. It was a good deal which I have continued.'

Just then an elderly gentleman came past using a walking stick.

'That reminds me of something that happened with Wolfgang. My husband broke his leg about four years ago and when he came off his crutches he was limping quite badly for a while. He started taking the dog for short walks and, believe it or not, Wolfgang slowed right down and started limping too.'

Pingu and Olivia

Back in the pet café and Ulla has fallen on her paws, as she put it, once again! She was there today to take pictures of André and Sebastian (André's Standard Poodle) as she thought that they were beginning to look very much like each other. While waiting for André to serve a customer, she began talking to Olivia and her French Bulldog, Pingu, who is a gorgeous steely grey with patches of white and looks at you as if she isn't quite sure what's going on.

'Thank you for introducing me to your lovely dog. Did you grow up with dogs at home?' asks Ulla to open the interview.

'Not all the time. We had our first family dog, a Labrador, when I was 8. When I was 18 I made a decision to get my own dog, a pug called Harley. My mother wanted nothing to do with Harley and told me to get rid of him, which of course I refused to do, so I left home. I have lived by myself for the last eight years and now I have little Pingu. She is totally amazing. She's crazy but she's amazing.'

'What made you choose this type of dog?'

'I really like her mushy, squished-up face. I'm not going to lie, she is a bit of a fashionable dog to have and I quite like that. They have incredible personalities.'

'You say you chose her, or did she choose you?'

'Weirdly enough, my best friend also wanted a French Bulldog and I went with her to see a litter. She wanted a fully blue one. When I saw Pingu she was the most incredible dog and I just had to have her. It was amazing.'

'She is certainly very sociable and seems very trusting. Where did you get her?'

'From a breeder in Streatham, you know, in South London. He was a proper breeder with all the Kennel Club stuff and everything. She's the most sociably interesting creature you'll ever meet in your entire life!'

'My next question is from Bryan. He's a scientist. I can tell you love Pingu, but can you tell me why? Bryan is curious to know if you can put your reasons into words.'

'That's easy. She gives me comfort without needing to talk. She reads me and knows what's wrong. I was going through a really bad time and she looked after me. She helped me through a lot and could sense when there was something wrong. She is becoming a lot like me.'

'I see what you mean.'

'People have said I looked like my Labrador, but I don't think I look like Pingu! I hope I'll never be old and wrinkly. But Pingu is as crazy as me. She is my baby. My absolute baby and she is completely reliant on me. It's like having a baby but not so much responsibility. I like to cuddle up with her but also to throw her around, gently of course. I actually quite like annoying her. Even my mum is in love with her.'

'It seems unnecessary to ask you whether you think of Pingu as one of the family.'

'She's my daughter. When my friend brings her dog round we talk to them like we were their parents. Like we say "Go to Mummy" and stuff.'

'Do you let Pingu sleep in your bed?'

'She sleeps in my bed, under the covers. She will come up, sit by my head, make sure I open the duvet so she can get under it. If I'm asleep, she wakes me up.'

'Does she cost you a lot of money to keep?'

'She gets all the best food and I'm paying a lot for life-long insurance, I think it's about £90 a month. My mother pays out a lot more for Harley, around £250 a month, which is crazy.'

'Do you find yourself buying presents for Pingu? Bryan and I went to Crufts, where we saw a village full of boutiques selling all sorts of goodies for dogs.'

'That's my dream. I want to go. She's my baby daughter and I don't see myself having any real daughters any time soon.'

'What about training Pingu?'

'She had basic training before I got her. Since then she's only had one training session but I didn't like the trainer, she was just too scary, like being with a scary teacher at school. So that didn't last long. Pingu's only problem is that if she sees a kid playing with a football she'll want to join in, but otherwise she's a really good girl.'

'Hopefully it is many years ahead, but when Pingu comes to the end of her life, would you ever think of cloning her?'

'Yes, I would. I'm going to have her spayed, but I would love to have her puppies! I could arrange for a surrogate and make babies with a very, very, very expensive dog and I'm going to get it done soon because I don't want to go another season with dogs following us everywhere. That was the worst thing that ever happened to me.'

'There is just one more thing to ask you. Bryan's book is about the evolution of dogs from wolves, so I wondered if you had any opinions about that?'

'I don't really understand it, to be honest. How are there so many different breeds when the whole point of evolution is mutations. So does that mean there were, like, French Bulldog wolves back in the day?'

'Hmmm.'

Battersea Dogs and Cats Home

Ulla's interviews show very clearly that the bond between owner and dog is extremely strong. No matter whether they owned a working dog or a pet, when asked the direct question, each replied that they loved their dog. But there is a dark side to this relationship. Thousands of dogs are abandoned every year. The Dogs Trust, a UK charity that re-homes abandoned dogs, estimated from local authority records that 47,000 dogs were given up in 2015 alone. In the USA, the Society for the Prevention of Cruelty to Animals reckons that 3.3 million dogs are taken into care centres every year. The reasons are many and various, so to find out more, Ulla and I arranged to visit Battersea Dogs and Cats Home, the oldest rescue centre in the world.

The Home was started by Mary Tealby in 1860 at the peak of the Victorian pet craze as the Temporary Home for Lost and Starving Dogs. It moved to its present premises in 1871, and in 1883 cats were admitted for the first time. The Home itself lies in the apex of a triangle formed between a major road and two railway lines just south of the River Thames. Dominating the skyline is the Art Deco bulk of Battersea Power Station, soon to disappear behind encircling blocks of luxury flats. Space on the site is at a premium, so there are two outstations in Windsor and Kent which deal with the overflow.

We entered through heavy steel gates, specially designed to prevent escapes, and found ourselves in a small open area surrounded by dogs on coloured leads being walked mostly by young ladies who, I discovered, were all volunteers. We were met by Hayley who, it is scarcely necessary to say, loved dogs and wished she could have one herself if only she had room. Battersea houses around 250 dogs and the same number of cats at any one time. We were soon surrounded by a cacophony of barks and yelps as Hayley guided us to the public holding cages. Here were the dogs that had been abandoned by their owners for one reason or another.

When they arrive at the Home, after a health check, the dogs are kept out of public view for the first few weeks to allow them to acclimatise to their new surroundings. Even dogs that have been well treated by their owners need time to settle in.

The atmosphere, though noisy, was surprisingly very relaxed. Volunteers sat playing with dogs in their pens. Calming music floated through the building. The dogs had their own blankets. Toys were scattered on the floor and murals decorated the walls. We saw Jack Russell Terriers, a beautiful Husky with piercing blue eyes, and several cross-breeds. Dogs arriving without a name are given one. Our visit coincided with the second week of the Wimbledon tennis championships. The Home, entirely funded as it is by charitable donations and ever alert to any PR opportunity, named the new arrivals after top players in the tournament – Roger the Jack Russell, Venus the Greyhound and Serena the West Highland Terrier.

Two breeds dominate the intake – Greyhounds no longer able to race, and Staffordshire Bull Terriers. Several charities specialise in finding homes for retired Greyhounds, but I wanted to hear why Staffies were over-represented. This breed was developed for fighting, and although dogfights are now illegal, owning a Staffie confers a certain macho status. They also have a reputation for being aggressive and difficult to handle, and a lot of abandoned Staffies were once owned by homeless people keeping them for protection. Bella was one of these. She came right up to the bars in her pen and looked straight at me, her head slightly tilted to one side. Her big pink mouth was open and her tongue was hanging out. One thought went through my mind: 'What have we done to this poor creature, bred for centuries to bite and maul, all for our cruel amusement?' Yet her expression forgave all these misdemeanours. She just wanted to be loved. Be reassured, dear reader, that this was the only time my emotions got the better of me.

The ultimate purpose of Battersea is of course to find homes for all these animals. All dogs are eventually housed and none is 'put to sleep'.

They arrive from a wide variety of sources, as Hayley explained. Among them are dogs whose owners can no longer provide a home. Perhaps they have moved and no longer have room, or their work has taken them abroad. Two delightful spaniels had arrived from Italy with owners who worked in the City but who were recalled to Italy almost straightaway. The dogs settled in well but had difficulty adjusting to the new healthy diet. Hayley explained that they were accustomed to eating only Parma ham, which would have stretched the feed budget of even the best-funded sanctuary!

We were guided out of the holding pens into a section of the Home not normally open to the public. Here was a play area for the dogs where they could be taken off their leads. There were more kennel blocks, one for dogs that had caught 'kennel cough', a contagious infection of the upper airways. Although it is treatable, once infected, dogs are never completely free of the virus so precautions are taken to segregate infected and 'clean' dogs. They are housed in different buildings, they have different-coloured leads, and as far as possible they never meet each other.

Close by was the reception area for new arrivals. This is accessed through a gate set back from the road so that owners can drop off their dogs without being seen. To say goodbye to a dog can be a very emotional experience for the owners, and there are many who change their minds at the last minute. A few owners are too embarrassed to bring their dog in and just leave them outside the gate.

Some dogs have evidently been physically maltreated, and Battersea has a team of canine behaviour specialists who deal with often unseen psychological abuse. These dogs may take a long time to settle but eventually they all do. In fact, the dogs quite soon forget all about their former lives. Some owners on the other hand never get over the trauma and possibly the guilt of giving up their dog for adoption, and ask to have it back. But once a dog has been re-housed it can never be reunited with its former owner. This might sound heartless, but the overriding consideration at Battersea is to do what is best for the dog.

On the one hand, the necessity of places like the Battersea Dogs and Cats Home is a consequence of human carelessness or cruelty towards dogs. On the other hand – and this came across to me most strongly – the centre and its staff are a testament to the enduring love between the two species that has grown up almost since time began.

24

Born Again: Cloning Your Dog

The interviews in the last chapter echo what we have seen throughout this book, that the emotional bond between dog and owner can be extremely strong. Dog owners know this of course, but I was surprised at the intensity of the responses. Without exception, all the owners professed to love their dog. When asked why, they were usually unable to say, I think because they'd never thought of it before. Nonetheless the phrase 'unconditional love' kept cropping up in the narratives. Most of the owners would be very sad if, or rather when, their dog died, but they were able to come to terms with the loss and often get another dog soon after. When it came to a replacement, a few owners would have considered cloning their pet, but most were put off by the price tag of $50,000 to $100,000. I suspect many more would go ahead if the cost fell to a similar level as a replacement, which is unlikely. There were also owners who would never contemplate cloning their dead pet, on what one might call ethical grounds. But what does cloning your dog actually involve? I decided to find out.

The death of a dog is always a devastating blow to the owner, sometimes almost as traumatic as the death of a child. But, unlike children, their relatively short lifespan means that the death of a pet is not an uncommon event. No surprise then that, following the birth of Dolly the sheep in 1996, the first animal cloned from an adult cell, bereft pet-owners around the world saw the opportunity of replacing their beloved pet with a genetically identical copy, and astute entrepreneurs sensed a golden business opportunity.

Cloning is the production of genetically identical individuals. It is the normal method of reproduction in many species of plant and animal that have jettisoned the arduous and inefficient business of sexual reproduction. Completely asexual species are ultimately doomed to extinction, of course, as they eventually succumb to parasites and pathogens that, having eventually unpicked the genetic defence of one individual, are able to sweep through the entire population. Being genetically identical, they are powerless to resist. Sex is the price we pay to avoid this threat by introducing genetic variability between individuals. All the same, asexual species are very successful – while they last. Among plants, dandelions and strawberries have abandoned sex, as have aphids (though they still indulge, briefly, at the end of the season) and some North American lizards.

Cloning is the norm for many agricultural and garden plants, like bananas, roses and fruit trees, where new plants are propagated through cuttings. There is nothing novel about cloning.

Scientists trying to unpick the secrets of embryonic development have taken advantage of the ability of some simple creatures, like flatworms and sea urchins, to regenerate new genetically identical individuals from 'cuttings' of adults. But this becomes increasingly difficult in more advanced organisms. It is completely impossible to grow an entire new mammal from a cutting, although the liver retains the ability to regenerate a new lobe without difficulty.

However, we are very familiar with what are literally human clones in the form of identical twins. Human twins are of two main types. Dizygotic, or non-identical twins, develop from separate fertilised eggs and are genetically no more alike than siblings. Identical or monozygotic twins grow from a single fertilised egg and are genetically identical clones.

The comparative abundance of both types of human twin has allowed scientists to study the extent to which any trait is influenced by genes on the one hand and, on the other, by the environment. If a trait is entirely determined by genetics, then 100 per cent of identical twins will share it. In other words, if one twin has the trait, whatever it is, so does the other in every case. This is termed complete concordance, but it is an entirely theoretical concept. In practice there are no traits which show 100 per cent concordance in identical twins. This is partly due to mutations that happen in one twin and not the other after the fertilised egg divides in two. A recent genetic analysis of blood cells in pairs of monozygotic twins detected differences in around 1 per 10 million DNA bases, leaving scope in a genome of 3,000 million bases for hundreds of individual mutations. The majority of these mutations will have no consequences at all, but if they disrupt vital genes there will be an effect on some body processes. Other differences between pairs of monozygotic twins are caused by so-called epigenetic effects. Still poorly understood, these are changes due to environmental differences experienced by the twins, either while they are developing in the womb, or after birth. Nonetheless, monozygotic twins can be astonishingly similar in both appearance and personality. While parents and siblings can always tell the difference between them, strangers usually cannot.

News of the great cloning breakthrough came when Ian Wilmut and his team at the Roslin Research Institute near Edinburgh published the results of their cloning experiments in 1997.[1] They had taken cells from the mammary gland of an adult blackface sheep and grown them on a cell culture dish for several days before withdrawing their nuclei with a fine glass tube and inserting them into sheep eggs from which the nucleus had been removed. Each of these steps resembled the manipulations required during in vitro fertilisation (IVF). The embryos created by this nuclear transfer were grown for a few more days before being transplanted to the uterus of a surrogate sheep. After a normal gestation two lambs were born, both apparently healthy.

This birth of healthy lambs was an undoubted triumph, especially as the nuclei had been taken from adult cells which many thought would not be able to give rise to all the cell types, like blood and bone for example, required to produce a lamb. However, success came at a cost. Three embryos were aborted, two lambs died within minutes of being born, and a third died after ten days. The whole process was clearly very wasteful.

The motivation for the experiments was to make it easier to produce identical transgenic sheep. Several years earlier, scientists at the Roslin Institute had manipulated sheep embryos which, when adults, were able to produce useful biological drugs, like clotting factors used to treat haemophilia, in their milk.

As soon as the news from Roslin broke, the possibilities for cloning other animals were being discussed openly. At first the focus was on agricultural animals, like cattle, where having identical individuals promised a revolution in husbandry. All cloned cattle would have the same milk yield, food conversion rates and temperament, thus all animals could be treated the same.

On the other side of the Atlantic it was not the prospects for animal husbandry that excited one particular resident in the Arizona desert. Billionaire John Sperling was relaxing one Sunday morning with his family at his luxurious mansion in the exclusive resort of Scottsdale, just east of the state capital Phoenix. As was customary, he spent his Sunday mornings scanning the numerous sections of *The New York Times*. On the front page of one was an article that caught his eye: under the headline 'Scientists Report First Cloning Ever of Adult Animal' was the news of the Roslin triumph. According to John Woestendiek, whose book *Dog, Inc.*[2] gives a very readable account of this and subsequent events, Sperling, looking at the dog curled up at his feet, immediately remarked, 'Hey, we should clone Missy.'

Sperling's wealth had accumulated through his private college, the University of Phoenix. In his later years he enjoyed using his money and his connections on eccentric projects, but always ones with a prospect of financial gain. It wasn't long before he, through his trusted lieutenant Lou Hawthorn, had contacted the world-renowned veterinary college Texas A&M University at Austin. Scientists at Texas A&M were well aware of Wilmut's work with Dolly the sheep and were thinking about how they could climb on board the cloning bandwagon. Whatever doubts there were about using dogs, rather than a species with a more obvious agricultural use, evaporated when Sperling offered $2.3 million to fund the work.

The research team at Texas A&M soon found that dogs were much harder to clone than sheep. The main technical difficulty was that the outer layer of a dog egg is unexpectedly opaque, making the necessary manipulations extremely difficult. Well aware that they were specifically funded to clone a dog and not a cat, the team asked for permission to switch in order to overcome the opacity of the dog egg. They would practise the other stages of the cloning process on cats before returning to work on dogs. This would cost more money than anticipated, but Sperling grudgingly agreed to the budget increase.

The scientists chose to clone a mottled cat with a coat made up of patches of black and white fur. Their choice was a curious one, which I suspect was an early sign of a developing tension between scientist and sponsor. The scientists were, I am quite sure, well aware that the coat-colour of cats is determined by a gene on the X-chromosome. As is widely known, females have two X-chromosomes while males have only one. There are a lot of genes on the X and mammals are very sensitive to the gene dosage, as it's called, so there is a potential problem. The usual fate of an embryo that has too many active genes is early abortion, so the question arises how this can be avoided in females. If every female cell had two active copies while all male cells had only one, there would be trouble.

The answer to this paradox was worked out by the British geneticist Mary Lyon and is embedded in all genetics courses. Lyon discovered that very early on in embryonic development, one of the X-chromosomes in each cell is switched off, and remains so in all the daughter cells created by cell division. However, the inactivation process is entirely random. Female mammals of all ages, including humans, are mosaics of cells, some with one X-chromosome in the active state and the rest with the other. Only rarely will anything show in the adult animal. However, crucially for the cat-cloning work, the coat-colour is decided by a gene on the X-chromosome. The first litter of cloned kittens certainly had black and white coats, but because of the random nature of X-inactivation, the patch pattern was quite different in each cat and different from the cat that was cloned.

The scientists learned what they needed to know about feline embryonic development, but the sponsors were furious. What was the use of cloning a pet if it came out looking different? Though nothing was ever proven, I can't help thinking that the scientists deliberately chose a mosaic donor cat to frustrate the commercial ambition of the single-minded sponsor. Such is the delicate balance between the academic and commercial worlds. The fracas led ultimately to the dissolution of the relationship between Texas A&M and the Sperling Foundation. I can just imagine the university's commercial arm foaming at the mouth.

Sperling and his associates persevered in their efforts but were overtaken by a team from the University of Seoul in South Korea, who found ways of overcoming the technical difficulties caused by the opacity of the dog eggs. It took several more years to produce the first cloned dog, but in 2005 Snuppy, an Afghan hound, was born. His name is an acronym derived from Seoul University Puppy Cloning. The cloning was still a very inefficient process. For instance, 1,095 transferred embryos produced only two live pups, of which Snuppy was one. Cloning is an expensive and controversial business that has struggled to become 'consumer-ready' in the words of Sperling's dog-cloning business, BioArts. Nonetheless, once they survive the first few days, cloned dogs appear to be free of systemic health issues.

In the paper announcing the birth of Snuppy, the lead scientist Byeong Lee predicted the use of cloning to immortalise elite or 'celebrity' dogs, for example those with an enhanced ability to detect cancers or those in the public eye. Trakr, a German Shepherd that had become a celebrity through his rescue of survivors of the terrorist attack on the World Trade Center in 2001, was cloned. He died in 2009, but his clones live on with their owner, James Symington, in Los Angeles.

The donor dog no longer has to be alive, so long as a piece of tissue or a blood sample has been carefully frozen. In 2012, a litter of five cloned Pitbull Terriers was born three years after the donor dog had died.

Dog cloning is still advertised, though no longer in the USA. It is flourishing in South Korea, where the man–dog relationship is rather different than in the West. Dog meat has been part of the diet there for at least 2,000 years, and a BBC documentary found that between 4,000 and 6,000

Trustt, Solace, Prodigy, Valor and Dejavu are two generations of cloned puppies of Trakr, a German Shepherd who sniffed out survivors from under the rubble of New York's World Trade Center.

restaurants were serving a type of dog soup. A friend of mine from Oxford, an expert on Korean languages and Korea in general, told me an amusing story which highlights the difference in the status of the dog in Korea and the West. During the 1988 Seoul Olympics there was a government clampdown on dog restaurants so as not to offend the sensitivities of Western visitors to the Games. Almost overnight, eateries that had once proudly displayed examples of their canine cuisine replaced the displays with inoffensive seafood and noodle dishes.

The centre for dog-cloning research moved to Seoul National University, Snuppy's alma mater, through the private spin-off Sooam Biotech Foundation. The foundation acquired a murky reputation when its founder Hwang Woo Suk announced to the world in 2004 that he had successfully cloned a human baby. He hadn't.

Given the deep emotional bond between man and dog, it is perhaps a surprise that cloning is not more popular. The cost is certainly one consideration, and another, not widely known, is that the whole process sacrifices the lives of many other dogs. The recipient eggs which are to receive the nuclear transfer are extruded from the ovaries of donor dogs, who do not always survive. The surrogate mothers are sometimes disposed of once the pups are sold on to their expectant owners. Meanwhile, owners expecting an exact replica of their beloved pet can be disappointed. They will certainly be similar but perhaps not exactly the same, just as no two identical human twins, as we have seen, are precisely alike in every respect.

It is not a surprise that cloned dogs are not identical in all respects; although they may look pretty much the same, they will behave differently. Just like the parent who has no difficulty in recognising which of their identical twins is which, the dog owner will be able to spot that the cloned puppy that 'arrives in the post' from Korea is not an identical replica of their beloved pet. And when it has cost a small fortune, the realisation that it isn't the same dog has led to disappointment and, more than once, a claim for a refund.

Even if the commercial market for cloned pets is limited or, in the words of John Sperling's now defunct BioArts, dog cloning is not yet 'consumer ready', I can imagine that cloning may well soon play a major role in the future evolution of the dog. After all, there is no reason to think that their evolution has stopped, and cloning might accelerate the propagation of unlikely mutants. Will we soon see the vegan dog or a miniature no bigger than a mobile phone,* or even a dog that, like the Hell Hound himself, glows in the dark?

* There have been recent news reports of the growing popularity of miniature 'teacup' dogs among minor celebrities and their followers on social media. Rather than being discarded, the 'runts' in a litter are chosen for breeding in the same way that freakish 'sports' were selected by Victorian breeders. Teacup dogs are crippled in various ways by their small size, but time will tell whether they can evolve to become viable.

25 Beyond the Reach of Reason

During the course of this book we have immersed ourselves deep into the detail of the canine genome. It tells us a great deal. We know that under the influence of artificial selection the wolf has been transformed by man into an animal with a vast range in size, colour, temperament and ability. We can even begin to follow these changes in molecular detail as we identify the genes, even the mutations themselves, which are responsible for size and coat-colour and other aspects of appearance. We have made a good start in identifying and developing tests for the genes responsible for many of the inherited disorders that plague pedigree dogs and which, given the will, could one day eliminate them altogether. We are beginning to explore the fascinating genetics behind canine behaviour. Why on earth do pointers 'point', for example? There must be a genetic explanation, probably quite a simple one, and one day we might discover the reason. All this is good, but to really understand why we love our dogs, we must look outside genetics.

Earlier in the book we saw how dogs and wolves were evaluated in behavioural testing and how they were certainly different from one another. These were rational tests on canines, but in this chapter I am going to explore a little further our reasons for loving these creatures. It has been suggested, not by dog lovers I am sure, that we have been manipulated by dogs.[1] The idea is not new. One of Aesop's fables, composed in second-century Greece, tells of a hungry wolf that after an unsuccessful day's hunting came across his domestic cousin, a well-fed mastiff. What do you have to do to be fed so well?' enquired the wolf. 'Very little,' replied the dog, 'just drive away beggars, guard the house, show fondness to the master and be submissive to the family.' The wolf considered what he had heard. Every day he risked his own life, he had to shelter from the worst of the weather and could never be sure of his next meal. 'If that is all you have to do, I think I'll give it a try. But tell me, what is that around your neck?' 'It's my collar for attaching my chain.' 'Your chain? You mean you are not free to go where you like?' 'Sure, but face the facts, it's only a small price to pay.' At this, the wolf turned and trotted back to the freedom of the forest.

The parasite theory of canine love proposes that dogs have somehow managed to burrow deep into our unconscious, identify our weaknesses and exploit them, much as a virus might do. Could dogs be the ultimate parasite, spared having to find food or shelter in return for merely putting on a cheap show of 'unconditional love'?[2] Already I hear the gasps of disapproval at this outrageous suggestion, but it remains a formal possibility for the hardwired Darwinian. Aesop clearly had it in mind in his fable of the wolf and the mastiff.

Something as strong, deep and mysterious as the bond between man and wolf is almost certain to be rooted in the unconscious mind – unknown territory to most scientists, including and perhaps even especially geneticists, whose explanations for the evolution of this amazing psychic symbiosis are prosaic in the extreme. No amount of scavenging human rubbish tips would account for the devotion and love that bind man and dog. What Shaun Ellis experienced in the forests of Idaho is the reciprocal nature of the bond. In his view he survived being eaten only because the wolves had something to gain from him. He imagined they saw him as a means of understanding the humans that hunted them. The first cooperative interaction between wolves and humans that I imagined in Chapter 1 had clear mutual benefits to both species.

The same is true of any symbiosis, any working relationship between different species. In Africa the honey guide, a relative of the woodpecker, invites humans to follow it to the nest of wild bees that the bird has earlier located. The man digs the nest from the hollow trunk and throws the bird part of it as a reward. Both man and bird gain from the interaction. The coastal fishermen of Laguna in Brazil cast their nets across shoals of sardines driven into the shallows by dolphins. The nets confine the fish, the dolphins take their pick and the fishermen go home with a much bigger haul than otherwise. In these two cases, the honey guide and the dolphin act as if they sense the intention of the human.

Our relationship with first the wolf and now the dog far exceeds the limited mental overlap required to get at the honey or the sardines, but nonetheless extends from the same principle of mutual benefit. With our overwhelming self-importance we might assume that we were the sole initiators of the 'domestication' of the wolf, that we were the only ones capable of realising the benefits of hunting together. We probably thought of it first, since it does not seem to have dawned on the Neanderthals. But I'm inclined to think that we greatly underestimate the part the wolf played in forging this relationship. The 'Great Awakening' of the Upper Palaeolithic might have given us the self-awareness to make the inventive step that started the process leading to cooperative hunting, but without the wolf's active participation nothing would have come of it. As it is, it must rank as one of the most important ingredients in ensuring first our survival and then our inexorable rise to our present dominance over all other creatures.

In one creation myth related by the psychoanalyst Patricia Dale-Green,[3] a gulf opened between Adam and the beasts he had named. Among the animals stood a dog. When the separation was almost complete the dog leaped across the gulf and took its place by the side of man.

The mythical significance of the bond between man and dog greatly interested Sigmund Freud, Carl Jung and many other psychoanalysts. A common theme is that both humans and dogs feel a strong need for attachment, for close interactions with others. This should not surprise us when we consider that both species depend on mutual cooperation for survival. The psychologist Edward Rees saw the human–dog bond as an example of basic reciprocal attachment, and instinct based on nurture, care giving and emotional and physical closeness.

The influential English psychologist John Bowlby fundamentally disagreed with Freud's view of instinct as learned behaviour in response to external influences, classically that of being fed. Bowlby developed the Jungian notion that instinct is part of an ancient archetype buried deep beneath our rational minds. Many of his patients' problems stemmed from the very common difficulty they had in integrating their rational minds with the unconscious archetypes that really pulled the strings. Dogs help to bridge this gap for some people, while for others leading lives lacking in attachment to other humans, isolation can lead to the obsessive care given to their pets. The dog becomes the object of transference and caring for it to excess represents their attempt to care for themselves.

Both man and wolf also share very well-developed family structures which have evolved over millennia. It may not always feel like it, but human family bonds are extraordinarily strong. An interesting psychological experiment carried out a few years ago on a group of students asked them for their responses to three hypothetical situations in which they needed to sacrifice something for either friends or family. The three situations were in ascending order of sacrifice: 1) lending emotional support, 2) lending money, and 3) donating a kidney. The respondents preferred to give emotional support to friends before family, they were pretty neutral when it came to lending money, but family won hands down when it came to donating a kidney. Even among students, blood, it seems, is still thicker than water *in extremis*.

The loyalty of dogs, so often commented on by owners in Ulla's interviews, stems directly from their unshakeable loyalty to the pack. The pack almost becomes a separate organism, rather like a beehive, whose survival is paramount even at the expense of individuals. In my imagined encounter near the Gate of Trajan in the first chapter, the alpha male threw himself at the enraged bull, mindless of his own safety in his desperation to save the she-wolf Lupa and preserve the integrity of the pack. In modern times, it is we who are objects of transference. We have become the pack. As we saw with the wolves of Longleat, the efforts to rebuild the social order within the pack after one of the pack members died were fraught with difficulties. Many dog trainers insist that humans must establish themselves as the dominant animal in the hybrid pack or there will be trouble. This insistence on dominance has led to much cruelty in training, but many people, including Paul Bradshaw, the author of *In Defence of Dogs*, reject the idea that cruelty is necessary.[4]

Both man and dog also share a highly developed ability for transference. The owners in Ulla's interviews considered their pet dogs to be members of the family. Some even referred to their pets in human terms, saying 'my baby' for example.

Many psychoanalysts, including Freud, practised their art in the presence of a dog. Eleora Woloy is one professional who has written about her experiences in a little book called the *Symbol of the Dog in the Human Psyche*.[5] The presence of her dog Tinsel, an old German Shepherd, relaxed her clients and helped them to reach into their unconscious minds. Tinsel could pick up signals of distress and would, without prompting, do something entirely appropriate. If a client was crying, Tinsel would lay her head on their lap.

Konrad Lorenz saw dogs as the mediators between humans and the natural world. Now that so many of us have severed our links with nature and experience only city life, the dog has become the last chance of retaining this most precious of connections.

Look deep into your dog's eyes and see, reflected there, distant crocus meadows shimmering beneath the snowy peaks of Carpathia.

Acknowledgements

This book owes everything to my two closest collaborators, my wife Ulla and her canine soulmate Sergio. In their different ways they forced me to revise my view of dogs from the dangerous, malodorous, drivelling wasters I once thought them to be, and to appreciate their empathetic and courageous nature.

My special thanks are due to the scientists, in particular Robert Wayne, Heidi Parker, Bridgett vonHoldt and Elaine Ostrander, who have together revealed the fascinating genetics on which this book is based. Many others deserve credit for their work, but these four stand out as genuine pioneers.

Others helped with particular sections, notably Ulla's army of willing volunteer owners and their dogs. The Kennel Club, especially its Librarian and Collections Manager Ciara Farrell, was tremendously helpful. Dr Catherine Mellersh from the Animal Health Trust patiently explained the practical application of modern genetics to canine health, while Hayley Chow from the Battersea Dogs and Cats Home showed us what can happen when mistreated dogs need rehoming. Shaun Ellis and his partner Kim introduced Ulla and me to their wolves at their farm in Cornwall, brought to life by Shaun's experiences with wild wolves in the forests of Idaho in the USA. Ethel Johnston and her husband John, along with Ewan Grant and Owen Macrae, enlightened me about the ways of working dogs on New Zealand's sheep stations.

All authors need help to coax their manuscript into life. Danielle Hobart from Saint Clair in Dunedin, New Zealand, enthusiastically transcribed Ulla's recorded interviews, while Robin Roberts-Gant refined her photographs. As always, my agent Luigi Bonomi and my editors Myles Archibald, Hazel Eriksson and Steve Dobell worked their magic.

References

Preface

1. Human and animal fossils have been found at Pestera cu Oase. See Trinkaus, Erik, Oana Moldovan, Stefan Milota, Adrian Bîlgăr, Laurenţiu Sarcina, Sheela Athreya, Shara E. Bailey, Ricardo Rodrigo, Gherase Mircea, Thomas Higham, Christopher Bronk Ramsey and Johannes van der Plicht. 2003. 'An Early Modern Human from the Peştera cu Oase, Romania.' *Proceedings of the National Academy of Science USA* 100, 11231 and Qiaomei Fu, Mateja Hajdinjak, Oana Teodora Moldovan, Silviu Constantin, Swapan Mallick, Pontus Skoglund, Nick Patterson, Nadin Rohland, Iosif Lazaridis, Birgit Nickel, Bence Viola, Kay Prüfer, Matthias Meyer, Janet Kelso, David Reich & Svante Pääbo. 2015. 'An Early Modern Human from Romania with a Recent Neanderthal Ancestor.' *Nature* 524, 216.
2. Lorentz, Konrad, 1954, *Man meets Dog*. Methuen, London.

Chapter 3: I Met a Traveller from an Antique Land

1. Cann, Rebecca L., Mark Stoneking and Allan C. Wilson. 1987. 'Mitochondrial DNA and Human Evolution.' *Nature* 325, 31.
2. Vilà, Carles, Peter Savolainen, Jesús E. Maldonado, Isabel R. Amorim, John E. Rice, Rodney L. Honeycutt, Keith A. Crandall,

Joakim Lundeberg and Robert K. Wayne. 1997. 'Multiple and Ancient Origins of the Domestic Dog.' *Science* 276, 1687.

3. Ibid.
4. Sundqvist, A-K, Susanne Björnerfeldt, Jennifer Leonard, Frank Hailer, Åke Hedhammar, H. Ellegren and Carles Vilà. 2006. 'Unequal Contribution of Sexes in the Origin of Dog Breeds.' *Genetics* 172, 1121.

Chapter 6: Let the Bones Speak

1. Germonpré, Mietje, Mikhail V. Sablin, Rhiannon E. Stevens, Robert E.M. Hedges, Michael Hofreiter, Mathias Stiller and Viviane R Després. 2009. 'Fossil Dogs and Wolves from Palaeolithic Sites in Belgium, the Ukraine and Russia: Osteometry, ancient DNA and stable isotopes.' *Journal of Archaeological Science* 36, 473.
2. Savolainen P., Y.P. Zhang, J. Luo, J. Lundeberg and T. Leitner. 2002. 'Genetic Evidence for an East Asian Origin of Domestic Dogs.' *Science*, 298, 1610.
3. Larson G., E.K. Karlsson, A. Perri, M.T. Webster, S.Y. Ho, J. Peters, P.W. Stahl, P.J. Piper, F. Lingaas, M. Fredholm, K.E. Comstock, J.F. Modiano, C. Schelling, A.I. Agoulnik, P.A. Leegwater, K. Dobney, J.D. Vigne, C. Vilà, L. Andersson, K. Lindblad-Toh. 2012. 'Rethinking Dog Domestication by Integrating Genetics, Archeology and Biogeography.' *Proceedings of the National Academy of Science USA* 109, 8878.
4. O. Thalmann, B. Shapiro, P. Cui, V.J. Schuenemann, S.K. Sawyer, D.L. Greenfield, M.B. Germonpré, M.V. Sablin, F. López-Giráldez, X. Domingo-Roura, H. Napierala, H-P. Uerpmann, D.M. Loponte, A.A. Acosta, L. Giemsch, R.W. Schmitz, B. Worthington, J.E. Buikstra, A. Druzhkova, A.S. Graphodatsky, N.D. Ovodov, N. Wahlberg, A.H. Freedman, R.M. Schweizer, K.P. Koepfli, J.A. Leonard, M. Meyer, J. Krause, S. Pääbo, R.E. Green, R.K. Wayne.

2013. 'Complete Mitochondrial Genomes of Ancient Canids Suggest a European Origin of Domestic Dogs.' *Science* 342, 871.

5. Ibid.

Chapter 8: Hunting with Wolves

1. Lorentz, Konrad, 1954, *Man meets Dog*. Methuen, London.
2. Shipman, Pat, 2015, *The Invaders*. Harvard University Press, Cambridge MA.

Chapter 9: Why Didn't Shaun Ellis Get Eaten by Wolves?

1. Ellis, Shaun and Penny Junor, 2010, *The Man who Lives with Wolves*. HarperCollins*Publishers*, London.
2. Mowat Farley, 1963, *Never Cry Wolf: The amazing true story of life among Arctic wolves*. McClelland and Stewart, Toronto.

Chapter 10: Friend or Foe?

1. Mowat Farley, 1963, *Never Cry Wolf: The amazing true story of life among Arctic wolves*. McClelland and Stewart, Toronto.

Chapter 11: A Touch of Evil

1. Lopez, Barry, 1978, *Of Wolves and Men*. Simon & Schuster, London.

Chapter 12: The Basic Framework

1. '"Super pack" of 400 wolves terrorise remote Russian town after killing 30 horses in just four days', *Daily Mail*, 7 February 2011. http://www.dailymail.co.uk/news/article-1354445/Super-pack-400-wolves-kill-30-horses-just-days-remote-Russian-village.html.
2. Bradshaw J, 2011, *In Defence of Dogs*. Allen Lane, London.
3. Murie, Adolph. 1944. 'The Wolves of Mount McKinley.' *Fauna of the National Parks of the United States* 5, 30.

Chapter 13: We See the First Dogs

1. Davis, S.J.M., Valla, F.R. 1978. 'Evidence for Domestication of the Dog 12,000 Years Ago in the Natufian's of Israel.' *Nature* 276, 608.
2. Columella, Lucius Junius Moderatus, 60 CE, *De Re Rustica*, Volume II: Book 7. Trans. E.S. Forster, Edward H. Heffner, 1954, *Loeb Classical Library* 407. Harvard University Press, Cambridge, MA.

Chapter 16: The Dog Genome

1. Lindblad-Toh K., C.M. Wade, T.S. Mikkelsen, E.K. Karlsson, D.B. Jaffe, M. Kamal, M. Clamp, J.L. Chang, E.J. Kulbokas, M.C. Zody, E. Mauceli, X. Xie, M. Breen, R.K. Wayne, E.A. Ostrander, C.P. Ponting, F. Galibert, D.R. Smith, P.J. DeJong, E. Kirkness, P. Alvarez, T. Biagi, W. Brockman, J. Butler, C.W. Chin, A. Cook, J. Cuff, M.J. Daly, D. DeCaprio, S. Gnerre, M. Grabherr, M. Kellis, M. Kleber, C. Bardeleben, L. Goodstadt, A. Heger, C. Hitte, L. Kim, K.P. Koepfli, H.G. Parker, J.P. Pollinger, S.M. Searle, N.B. Sutter, R. Thomas, C. Webber, J. Baldwin, A. Abebe, A. Abouelleil, L. Aftuck, M. Ait-Zahra, T. Aldredge, N. Allen, P. An, S. Anderson, C. Antoine, H. Arachchi, A. Aslam, L. Ayotte, P. Bachantsang, A. Barry, T. Bayul, M. Benamara, A. Berlin, D. Bessette, B. Blitshteyn, T. Bloom, J. Blye, L. Boguslavskiy, C. Bonnet, B. Boukhgalter, A. Brown, P. Cahill, N. Calixte, J. Camarata, Y. Cheshatsang, J. Chu, M. Citroen, A. Collymore, P. Cooke, T. Dawoe, R. Daza, K. Decktor, S. DeGray, N. Dhargay, K. Dooley, P. Dorje, K. Dorjee, L. Dorris, N. Duffey, A. Dupes, O. Egbiremolen, R. Elong, J. Falk, A. Farina, S. Faro, D. Ferguson, P. Ferreira, S. Fisher, M. FitzGerald, K. Foley, C. Foley, A. Franke, D. Friedrich, D. Gage, M. Garber, G. Gearin, G. Giannoukos, T. Goode, A. Goyette, J. Graham, E. Grandbois, K. Gyaltsen, N. Hafez, D. Hagopian, B. Hagos, J. Hall, C. Healy, R. Hegarty, T. Honan, A. Horn, N. Houde, L. Hughes, L. Hunnicutt, M. Husby, B. Jester, C. Jones, A. Kamat, B. Kanga, C.

Kells, D. Khazanovich, A.C. Kieu, P. Kisner, M. Kumar, K. Lance, T. Landers, M. Lara, W. Lee, J.P. Leger, N. Lennon, L. Leuper, S. LeVine, J. Liu, X. Liu, Y. Lokyitsang, T. Lokyitsang, A. Lui, J. Macdonald, J. Major, R. Marabella, K. Maru, C. Matthews, S. McDonough, T. Mehta, J. Meldrim, A. Melnikov, L. Meneus, A. Mihalev, T. Mihova, K. Miller, R. Mittelman, V. Mlenga, L. Mulrain, G. Munson, A. Navidi, J. Naylor, T. Nguyen, N. Nguyen, C. Nguyen, T. Nguyen, R. Nicol, N. Norbu, C. Norbu, N. Novod, T. Nyima, P. Olandt, B. O'Neill, K. O'Neill, S. Osman, L. Oyono, C. Patti, D. Perrin, P. Phunkhang, F. Pierre, M. Priest, A. Rachupka, S. Raghuraman, R. Rameau, V. Ray, C. Raymond, F. Rege, C. Rise, J. Rogers, P. Rogov, J. Sahalie, S. Settipalli, T. Sharpe, T. Shea, M. Sheehan, N. Sherpa, J. Shi, D. Shih, J. Sloan, C. Smith, T. Sparrow, J. Stalker, N. Stange-Thomann, S. Stavropoulos, C. Stone, S. Stone, S. Sykes, P. Tchuinga, P. Tenzing, S. Tesfaye, D. Thoulutsang, Y. Thoulutsang, K. Topham, I. Topping, T. Tsamla, H. Vassiliev, V. Venkataraman, A. Vo, T. Wangchuk, T. Wangdi, M. Weiand, J. Wilkinson, A. Wilson, S. Yadav, S. Yang, X. Yang, G. Young, Q. Yu, J. Zainoun, L. Zembek, A. Zimmer and E.S. Lander. 2005. 'Genome Sequence, Comparative Analysis and Haplotype Structure of the Domestic Dog.' *Nature* 438, 803.

Chapter 17: The Genetics of Pedigree Breeds

1. vonHoldt, Bridgett M., John P. Pollinger, Kirk E. Lohmueller, Eunjung Han, Heidi G. Parker, Pascale Quignon, Jeremiah D. Degenhardt, Adam Boyko, Dent A. Earl, Adam Auton, Andy Reynolds, Kasia Bryc, Abra Brisbin, James C. Knowles, Dana S. Mosher, Tyrone C. Spady, Abdel Elkahloun, Eli Geffen, Malgorzata Pilot, Wlodzimierz Jedrzejewski, Claudia Greco, Ettore Randi, Danika Bannasch, Alan Wilton, Jeremy Shearman, Marco Musiani, Michelle Cargill, Paul G. Jones, Zuwei Qian, Wei Huang, Zhao-Li

Ding, Ya-ping Zhang, Carlos D. Bustamante, Elaine A. Ostrander, John Novembre, and Robert K. Wayne. 2010. 'Genome-wide SNP and Haplotype Analyses Reveal a Rich History Underlying Dog Domestication.' *Nature* 464, 898.

2. Parker Heidi G., Lisa V. Kim, Nathan B. Sutter, Scott Carlson, Travis D. Lorentzen, Tiffany B. Malek, Gary S. Johnson, Hawkins B. DeFrance, Elaine A. Ostrander and Leonid Kruglyak. 2004. 'Genetic Structure of the Purebred Domestic Dog.' *Science* 304, 1160.

Chapter 18: The Dance of Life

1. Parker, Heidi G., Dayna L. Dreger, Maud Rimbault, Brian W. Davis, Alexandra B. Mullen, Gretchen Carpintero-Ramirez, Elaine A. Ostrander. 2017. 'Genomic Analyses Reveal the Influence of Geographic Origin, Migration, and Hybridization on Modern Dog Breed Development.' *Cell Reports* 19, 697.

Chapter 19: At the Heart of the Matter

1. Eliot, T.S., 1939, *Old Possum's Book of Practical Cats*. Faber & Faber, London.
2. Boyle, Evan A., Yang I. Li and Jonathan K. Pritchard. 2015. 'An Expanded View of Complex Traits: From polygenic to omnigenic.' *Cell* 169, 1177.
3. Chase, Kevin, David R. Carrier, Frederick R. Adler, Tyler Jarvik, Elaine A. Ostrander, Travis D. Lorentzen and Karl G. Lark. 2002. 'Genetic Basis for Systems of Skeletal Quantitative Traits: Principal component analysis of the canid skeleton.' *Proceedings of the National Academy of Science USA* 99, 9930.
4. Sutter N.B., C.D. Bustamante, K. Chase, M.M. Gray, K. Zhao, L. Zhu, B. Padhukasahasram, E. Karlins, S. Davis, P.G. Jones, P. Quignon, G.S. Johnson, H.G. Parker, N. Fretwell, D.S. Mosher, D.F. Lawler, E. Satyaraj, M. Nordborg, K.G. Lark, R.K. Wayne, E.A.

Ostrander. 2007. 'A Single IGF1 Allele is a Major Determinant of Small Size in Dogs.' *Science* 316, 112.

5. Parker, Heidi G., Dayna L. Dreger, Maud Rimbault, Brian W. Davis, Alexandra B. Mullen, Gretchen Carpintero-Ramirez, Elaine A. Ostrander. 2017. 'Genomic Analyses Reveal the Influence of Geographic Origin, Migration, and Hybridization on Modern Dog Breed Development.' *Cell Reports* 19, 697.
6. Schoenebeck. J.J., E.A. Ostrander. 2013. 'The Genetics of Canine Skull Shape Variation.' *Genetics* 193, 317.
7. Young, Amy E., Jeanne R. Ryun, Danika L. Bannasch. 2006. 'Deletions in the COL10A1 Gene are not Associated with Skeletal Changes in Dogs.' *Mammalian Genome* 17, 761.
8. Schuelke M., K.R. Wagner, L.E. Stolz, C. Hübner, T. Riebel, W. Kömen, T. Braun, J.F. Tobin, S.J. Lee. 2004. 'Myostatin Mutation Associated with Gross Muscle Hypertrophy in a Child.' *New England Journal of Medicine* 350, 2682.
9. Mosher, Dana S., Pascale Quignon, Carlos D. Bustamante, Nathan B. Sutter, Cathryn S. Mellersh, Heidi G. Parker, Elaine A. Ostrander. 2007. 'A Mutation in the Myostatin Gene Increases Muscle Mass and Enhances Racing Performance in Heterozygote Dogs.' *PLOS Genetics* 5, e79.
10. Kambadur R., M. Sharma, T.P. Smith, J.J. Bass. 1997. 'Mutations in Myostatin (GDF8) in Double-Muscled Belgian Blue and Piedmontese Cattle.' *Genome Research* 7, 910.
11. Bannasch, Danika, Noa Safra, Amy Young, Nili Karmi, R.S. Schaible, G.V. Ling. 2008. 'Mutations in the SLC2A9 Gene Cause Hyperuricosuria and Hyperuricemia in the Dog.' *PLOS Genetics* 4, e1000246.
12. Hillbertz, Salmon N.H., M. Isaksson, E.K. Karlsson, E. Hellmén, G.R. Pielberg, P. Savolainen, C.M. Wade, H. von Euler, U. Gustafson, A. Hedhammar, M. Nilsson, K. Lindblad-Toh, L.

Andersson, G. Andersson. 2007. 'Duplication of FGF3, FGF4, FGF19 and ORAOV1 Causes Hair Ridge and Predisposition to Dermoid Sinus in Ridgeback Dogs.' *Nature Genetics* 39, 1318.

13. Leegwater, Peter A., Marjan A. van Hagen and Bernard A. van Oost. 2007. 'Localization of White Spotting Locus in Boxer Dogs on CFA20 by Genome-Wide Linkage Analysis with 1500 SNPs.' *Journal of Heredity* 98, 549.
14. Cadieu, Edouard, Mark W. Neff, Pascale Quignon, Kari Walsh, Kevin Chase, Heidi G. Parker, Bridgett M. vonHoldt, Alison Rhue, Adam Boyko, Alexandra Byers, Aaron Wong, Dana S. Mosher, Abdel G. Elkahloun, Tyrone C. Spady, Catherine André, K. Gordon Lark, Michelle Cargill, Carlos D. Bustamante, Robert K. Wayne and Elaine A. Ostrander. 2009. 'Coat Variation in the Domestic Dog Is Governed by Variants in Three Genes.' *Science* 326, 150.
15. Lin, L, J. Faraco, R. Li, H. Kadotani, W. Rogers, X. Lin, X. Qiu, P.J. de Jong, S. Nishino and E. Mignot. 1999. 'The Sleep Disorder Canine Narcolepsy is Caused by a Mutation in the Hypocretin (Orexin) Receptor 2 Gene.' *Cell* 98, 365.
16. Ibid.
17. Williams, J. 1961. *Circulation* XXIV, 1311.
18. vonHoldt, Bridgett M., Emily Shuldiner, Ilana Janowitz Koch, Rebecca Y. Kartzinel, Andrew Hogan, Lauren Brubaker, Shelby Wanser, Daniel Stahler, Clive D. L. Wynne, Elaine A. Ostrander, Janet S. Sinsheimer and Monique A. R. Udell. 2017. 'Structural Variants in Genes Associated with Human Williams-Beuren Syndrome Underlie Stereotypical Hypersociability in Domestic Dogs.' *Science Advances* 3, e1700398.

Chapter 20: In the Lab

1. Mellersh, Cathryn. 2012. 'DNA Testing and Domestic Dogs.' *Mammalian Genome* 23, 109

2. Forman Oliver P., Luisa De Risio and Cathryn Mellersh. 2013. 'Missense Mutation in CAPN1 Is Associated with Spinocerebellar Ataxia in the Parson Russell Terrier Dog Breed.' *PLOS ONE* 8, e64627.

Chapter 21: The Scientist Who Came in from the Cold

1. Nagasawa, Miho, Shouhei Mitsui, Shiori En, Nobuyo Ohtani, Mitsuaki Ohta, Yasuo Sakuma, Tatsushi Onaka, Kazutaka Mogi and Takefumi Kikusui. 'Oxytocin-Gaze Positive Loop and the Coevolution of Human–Dog Bonds.' *Science* 348, 333.

Chapter 24: Born Again: Cloning Your Dog

1. Wilmut, I., A.E. Schnieke, J. McWhir, A.J. Kind and K.H.S. Campbell. 1997. 'Viable Offspring Derived from Fetal and Adult Mammalian Cells.' *Nature* 385, 810.
2. Woestendiek, John, 2012, *Dog, Inc.: How a collection of visionaries, rebels, eccentrics, and their pets launched the commercial dog cloning industry*. Avery Publishing Group, New York, NY.

Chapter 25: Beyond the Reach of Reason

1. Schleidt, Wolfgang and Shalter, M.D. 2003. 'Co-evolution of Humans and Canids.' *Evolution and Cognition* 9, 57.
2. Ibid.
3. Dale-Green, Patricia, 1966, *Dog*. Hart-Davis, London.
4. Bradshaw, Paul, 2012, *In Defence of Dogs: Why dogs need our understanding*. Penguin, London.
5. Woloy, Eleanor, 1990, *The Symbol of the Dog in the Human Psyche: A study of the human–dog bond*. Chiron, Wilmette, IL.

Picture Credits

Page v: Illustration courtesy of Richard Sykes. This illustration depicts the tomb of Liliana Crociati de Szaszak in La Recoleta Cemetery, Buenos Aires, Argentina, which is known for its unusual neo-gothic design. Liliana was twenty-six years old when she was killed by an avalanche, and after his death several years later, her beloved dog, Sabú was added to her memorial. The text under the dog's statue reads 'Sabú, faithful friend of Liliana'.

Page 8: © GraphicaArtis

Page 11: Science History Images/Alamy Stock Photo

Page 17: Images courtesy of Professor Bryan Sykes.

Page 42: Getty/AFP/Staff

Page 43: JAVIER TRUEBA/MSF/SCIENCE PHOTO LIBRARY

Page 48: KENNIS AND KENNIS/MSF/SCIENCE PHOTO LIBRARY

Page 54: The Royal Belgian Institute of Natural Science, Wilfrid Miseur.

Page 81: Jose Schell/Nature Picture Library

Page 89: INTERFOTO/Alamy Stock Photo

Page 117: vonHoldt, Bridgett M., John P. Pollinger, Kirk E. Lohmueller, Eunjung Han, Heidi G. Parker, Pascale Quignon, Jeremiah D. Degenhardt, Adam Boyko, Dent A. Earl, Adam Auton, Andy Reynolds, Kasia Bryc, Abra Brisbin, James C. Knowles, Dana S. Mosher, Tyrone C. Spady, Abdel Elkahloun, Eli Geffen, Malgorzata

Pilot, Wlodzimierz Jedrzejewski, Claudia Greco, Ettore Randi, Danika Bannasch, Alan Wilton, Jeremy Shearman, Marco Musiani, Michelle Cargill, Paul G. Jones, Zuwei Qian, Wei Huang, Zhao-Li Ding, Ya-ping Zhang, Carlos D. Bustamante, Elaine A. Ostrander, John Novembre, and Robert K. Wayne. 2010. 'Genome-wide SNP and Haplotype Analyses Reveal a Rich History Underlying Dog Domestication.' Nature 464, 898. Reprinted by permission from *Springer Nature*.

Page 153: Getty / Donald M. Jones / Minden Pictures

Page 173: SPUTNIK / SCIENCE PHOTO LIBRARY

Page 186: Image courtesy of Ulla Plougmand

Page 198: Image courtesy of Ulla Plougmand

Page 213: Image courtesy of Ulla Plougmand

Page 229: Image courtesy of Phodography UK – Ursula Aitchison

Page 252: Getty / GABRIEL BOUYS / Staff

Index

Page references in *italics* indicate images.

INDEX

INDEX